Lecture Notes in Artificial Intelligence 5655

Edited by R. Goebel, J. Siekmann, and W. Wahlster

Subseries of Lecture Notes in Computer Science

Angelo Oddi   François Fages
Francesca Rossi (Eds.)

Recent Advances in Constraints

13th Annual ERCIM International Workshop
on Constraint Solving
and Constraint Logic Programming, CSCLP 2008
Rome, Italy, June 18-20, 2008
Revised Selected Papers

 Springer

Series Editors

Randy Goebel, University of Alberta, Edmonton, Canada
Jörg Siekmann, University of Saarland, Saarbrücken, Germany
Wolfgang Wahlster, DFKI and University of Saarland, Saarbrücken, Germany

Volume Editors

Angelo Oddi
ISTC-CNR, Institute of Cognitive Science and Technology
National Research Council of Italy
Via San Martino della Battaglia 44, 00185 Rome, Italy
E-mail: angelo.oddi@istc.cnr.it

François Fages
INRIA Paris-Rocquencourt
BP 105, 78153 Le Chesnay Cedex, France
E-mail: francois.fages@inria.fr

Francesca Rossi
University of Padova, Department of Pure and Applied Mathematics
Via Trieste 63, 35121 Padova, Italy
E-mail: frossi@math.unipd.it

Library of Congress Control Number: 2009930948

CR Subject Classification (1998): I.2.3, F.3.1-2, F.4.1, D.3.3, F.2.2, G.1.6, I.2.8

LNCS Sublibrary: SL 7 – Artificial Intelligence

ISSN   0302-9743
ISBN-10   3-642-03250-8 Springer Berlin Heidelberg New York
ISBN-13   978-3-642-03250-9 Springer Berlin Heidelberg New York

springer.com

© Springer-Verlag Berlin Heidelberg 2009
Printed in Germany

Typesetting: Camera-ready by author, data conversion by Scientific Publishing Services, Chennai, India
Printed on acid-free paper   SPIN: 12719471   06/3180   5 4 3 2 1 0

Preface

Constraint programming (CP) is a powerful programming paradigm for the declarative description and the effective solving of large combinatorial problems. Based on a strong theoretical foundation, it is increasingly attracting commercial interest. Since the 1990s, CP has been deployed by many industry leaders, in particular to model heterogeneous optimization and satisfaction problems. Examples of application domains where such problems naturally arise, and where constraint programming has made a valuable contribution, are scheduling, production planning, communication networks, routing, planning of satellite missions, robotics, and bioinformatics.

This volume contains the papers selected for the post-proceedings of the 13th International Workshop on Constraint Solving and Constraint Logic Programming (CSCLP 2008) held during June 18–20, 2008 in Rome, Italy. This workshop was organized as the 13th meeting of the working group on Constraints of the European Research Consortium for Informatics and Mathematics (ERCIM), continuing a series of workshops organized since the creation of the working group in 1997. A selection of papers of these annual workshops plus some additional contributions have been published since 2002 in a series of volumes which illustrate the evolution in the field, under the title "Recent Advances in Constraints" in the *Lecture Notes in Artificial Intelligence* series.

This year there were 14 submissions to the proceedings. Each submission was reviewed by three reviewers and the Program Committee decided to accept nine papers for publication in this volume. The papers in this volume present original research results, as well as applications, in many aspects of constraint solving and constraint logic programming. Research topics that can be found in the papers are first-order constraints, symmetry breaking, global constraints, constraint optimization problems, distributed constraint solving problems, soft constraints, as well as the analysis of application domains such as cumulative resource problems and hybrid systems.

The editors would like to take the opportunity to thank all the authors who submitted a paper to this volume, as well as the reviewers for their helpful and invaluable work. The organization of the CSCLP 2008 workshop and the publication of this volume was made possible thanks to the support of the European Research Consortium for Informatics and Mathematics (ERCIM), the Planning and Scheduling Team (PST) at the Institute of Cognitive Science and Technology (ISTC-CNR) of Rome, the Association for Constraint Programming (ACP), and the Department of Pure and Applied Mathematics, University of Padova, Italy. We hope that the present volume is useful to everyone interested in the recent advances and trends in constraint programming, constraint solving, problem modeling, and applications.

May 2009

Angelo Oddi
François Fages
Francesca Rossi

Organization

CSCLP 2008 was organized by the ERCIM Working Group on Constraints.

Organizing and Program Committee

Angelo Oddi	ISTC-CNR, Rome, Italy
Amedeo Cesta	ISTC-CNR, Rome, Italy
François Fages	INRIA, Rocquencourt, France
Nicola Policella	ESA-ESOC, Darmstadt, Germany
Francesca Rossi	University of Padova, Italy

Additional Reviewers

Olivier Bailleux	Barry O'Sullivan
Thanasis Balafoutis	Riccardo Rasconi
Stefano Bistarelli	Igor Razgon
Andreas Eggers	Andrea Rendl
Ian Gent	Francesco Santini
Giorgio Gosti	Kostas Stergiou
Natalia Kalinnik	Tino Teige
George Katsirelos	Kristen Brent Venable
Stefan Kupferschmid	Toby Walsh
Julien Martin	William Yeoh
Ian Miguel	Khalil Djelloul
Nina Narodytska	

Local Organization

Gabriella Cortellessa	ISTC-CNR, Rome, Italy
Riccardo Rasconi	ISTC-CNR, Rome, Italy

Sponsoring Institutions

European Research Consortium for Informatics and Mathematics (ERCIM)
Planning and Scheduling Team (PST)
Institute of Cognitive Science and Technology (ISTC-CNR) of Rome
The Association for Constraint Programming (ACP)
Department of Pure and Applied Mathematics, University of Padova, Italy

Table of Contents

From Marriages to Coalitions: A Soft CSP Approach

Stefano Bistarelli[1,2,3], Simon Foley[4], Barry O'Sullivan[4,5],
and Francesco Santini[1,2,6]

[1] Dipartimento di Scienze, Università "G. d'Annunzio" di Chieti-Pescara, Italy
{bista,santini}@sci.unich.it
[2] Istituto di Informatica e Telematica (CNR), Pisa, Italy
{stefano.bistarelli,francesco.santini}@iit.cnr.it
[3] Dipartimento di Matematica e Informatica, Università di Perugia, Italy
bista@dipmat.unipg.it
[4] Department of Computer Science, University College Cork, Ireland
{s.foley,b.osullivan}@cs.ucc.ie
[5] Cork Constraint Computation Centre, University College Cork, Ireland
b.osullivan@4c.ucc.ie
[6] IMT - Scuola di Studi Avanzati, Lucca, Italy
f.santini@imtlucca.it

Abstract. In this work we represent the *Optimal Stable Marriage* problem as a *Soft Constraint Satisfaction Problem*. In addition, we extend this problem from couples of individuals to coalitions of generic agents, in order to define new coalition-formation principles and stability conditions. In the coalition case, we suppose the preference value as a trust score, since trust can describe the belief of a node in the capabilities of another node, in its honesty and reliability. Semiring-based soft constraints represent a general and expressive framework that is able to deal with distinct concepts of optimality by only changing the related c-semiring structure, instead of using different ad-hoc algorithms. At last, we propose an implementation of the classical OSM problem using integer linear programming tools.

1 Introduction

The *Stable Marriage* (SM) problem [13,19] and its many variants [16] have been widely studied in the literature, because of the inherent appeal of the problem and its important practical applications. A classical instance of the problem comprises a bipartite set of n men and n women, and each person has a preference list in which they rank all members of the opposite sex in a strict total order. Then, a match MT is simply a bijection between men and women. A man m_i and a woman w_j form a *blocking pair* for MT if m_i prefers w_j to his partner in MT and w_j prefers m_i to her partner in MT. A matching that involves no blocking pair is said to be *stable*, otherwise the matching is unstable. Even though the SM problem has its roots as a combinatorial problem, it has also been studied in game theory, economics and in operations research [10].

A. Oddi, F. Fages, and F. Rossi (Eds.): CSCLP 2008, LNAI 5655, pp. 1–15, 2009.

However, in this paper we mainly concentrate on its optimization version, the *Optimal Stable Marriage* (OSM) problem [16,19], which tries to find a match that is not only stable, but also "good" according to some criterion based on the preferences of all the individuals. Classical solutions deal instead only with men-optimal (or women-optimal) marriages, in which every man (woman), gets his (her) best possible partner.

We propose soft constraints as a very expressive framework where it is possible to cast different kinds of optimization criteria by only modifying the c-semiring [1,4] structure on which the corresponding *Soft Constraint Satisfaction Problem* (SCSP) [1] is based. In this sense, soft constraints prove to be a more general solving framework with respect to the other ad-hoc algorithms presented in literature for each different optimization problem [16]. In fact, we can also deal with problem extensions such as incomplete preference lists and ties in the same list. Therefore, in this paper we build a bridge between the OSM problems and soft constraint satisfaction, as previously done between SM and classic constraint satisfaction [10,23]. Moreover, we use integer linear programming (ILP) as a general method to solve these problems. The classical SM problem (thus, the non-optimal version of the problem) has been already studied and solved by using crisp constraints in [10,23]. In [10] the authors present two different encodings of an instance of SM as an instance of a *constraint satisfaction problem* (CSP). Moreover, they show that *arc consistency* propagation achieves the same results as the classical *Extended Gale/Shapley* (EGS) algorithm, thus easily deriving the men/women-optimal solution [10].

The second main result provided in the paper relates to extending the stable marriage definition from pairs of individuals to coalitions of agents. A coalition can be defined as a temporary alliance among agents, during which they co-operate in joint action for a common task [14]. Moreover, we use trust scores instead of plain preferences in order to evaluate the relationships among agents. Therefore, the notion of SM stability is translated to coalitions, and the problem is still solved by exploiting the optimization point of view: the final set of coalitions is stable and is the most trustworthy with respect to the used trust metric, represented by a c-semiring [2,6,22]. Even for this coalition extension we use soft constraints to naturally model the problem.

The remainder of this paper is organized as follows. In Section 2 we summarize the background on soft constraints, while Section 3 does the same for the OSM problem. In Section 4 we represent the OSM problem with soft constraints and we solve it with ILP. Section 5 extends the OSM problem to coalitions, still representing the problem with soft constraints. Finally, Section 6 presents our conclusions and directions for future work.

2   Soft Constraints

A c-semiring [1,4] S (or simply semiring in the following) is a tuple $\langle A, +, \times, \mathbf{0}, \mathbf{1} \rangle$ where A is a set with two special elements $(\mathbf{0}, \mathbf{1} \in A)$ and with two operations $+$ and $\times$ that satisfy certain properties: $+$ is defined over (possibly infinite) sets of elements of A and thus is commutative, associative, idempotent, it is closed and $\mathbf{0}$ is its unit element and $\mathbf{1}$ is its absorbing element; $\times$ is closed, associative, commutative, distributes over $+$, $\mathbf{1}$ is its unit element, and $\mathbf{0}$ is its absorbing element (for the exhaustive definition, please refer to [4]). The $+$ operation defines a partial order $\leq_S$ over A such that $a \leq_S b$ iff $a + b = b$; we say that $a \leq_S b$ if b represents a value *better* than a. Other properties related to the two operations are that $+$ and $\times$ are monotone on $\leq_S$, $\mathbf{0}$ is its minimum and $\mathbf{1}$ its maximum, $\langle A, \leq_S \rangle$ is a complete lattice and $+$ is its lub. Finally, if $\times$ is idempotent, then $+$ distributes over $\times$, $\langle A, \leq_S \rangle$ is a complete distributive lattice and $\times$ its glb.

A *soft constraint* [1,4] may be seen as a constraint where each instantiation of its variables has an associated preference. Given $S = \langle A, +, \times, \mathbf{0}, \mathbf{1} \rangle$ and an ordered set of variables V over a finite domain D, a soft constraint is a function which, given an assignment $\eta : V \to D$ of the variables, returns a value of the semiring. Using this notation $\mathcal{C} = \eta \to A$ is the set of all possible constraints that can be built starting from S, D and V. Any function in $\mathcal{C}$ involves all the variables in V, but we impose that it depends on the assignment of only a finite subset of them. So, for instance, a binary constraint $c_{x,y}$ over variables x and y, is a function $c_{x,y} : V \to D \to A$, but it depends only on the assignment of variables $\{x, y\} \subseteq V$ (the *support* of the constraint, or *scope*). Note that $c\eta[v := d_1]$ means $c\eta'$ where η' is η modified with the assignment $v := d_1$. Note also that $c\eta$ is the application of a constraint function $c : V \to D \to A$ to a function $\eta : V \to D$; what we obtain, is a semiring value $c\eta = a$. $\bar{\mathbf{0}}$ and $\bar{\mathbf{1}}$ respectively represent the constraint functions associating $\mathbf{0}$ and $\mathbf{1}$ to all assignments of domain values; in general, the $\bar{a}$ function returns the semiring value a.

Given the set $\mathcal{C}$, the combination function $\otimes : \mathcal{C} \times \mathcal{C} \to \mathcal{C}$ is defined as $(c_1 \otimes c_2)\eta = c_1\eta \times c_2\eta$ (see also [1,4]). Informally, performing the $\otimes$ or between two constraints means building a new constraint whose support involves all the variables of the original ones, and which associates with each tuple of domain values for such variables a semiring element which is obtained by multiplying the elements associated by the original constraints to the appropriate sub-tuples. The partial order $\leq_S$ over $\mathcal{C}$ can be easily extended among constraints by defining $c_1 \sqsubseteq c_2 \iff c_1\eta \leq c_2\eta$. Consider the set $\mathcal{C}$ and the partial order $\sqsubseteq$. Then an entailment relation $\vdash \subseteq \wp(\mathcal{C}) \times \mathcal{C}$ is defined such that for each $C \in \wp(\mathcal{C})$ and $c \in \mathcal{C}$, we have $C \vdash c \iff \bigotimes C \sqsubseteq c$ (see also [1]).

Given a constraint $c \in \mathcal{C}$ and a variable $v \in V$, the *projection* [1,3,4] of c over $V - \{v\}$, written $c \Downarrow_{(V \setminus \{v\})}$ is the constraint c' such that $c'\eta = \sum_{d \in D} c\eta[v := d]$. Informally, projecting means eliminating some variables from the support.

A SCSP [1] defined as $P = \langle C, con \rangle$ (C is the set of constraints and $con \subseteq V$, i.e. a subset the problem variables). A problem P is α-consistent if $blevel(P) = \alpha$ [1]; P is instead simply "consistent" iff there exists $\alpha >_S \mathbf{0}$ such that P is α-consistent [1]. P is inconsistent if it is not consistent. The *best level of consistency* notion defined as $blevel(P) = Sol(P) \Downarrow_\emptyset$, where $Sol(P) = (\bigotimes C) \Downarrow_{con}$ [1].

3   The Optimal Stable Marriage Problem

An instance of the classical stable marriage problem (SM) [9] comprises n men and n women, and each person has a preference list in which all members of the opposite sex are ranked in a strict total order. All men and women must be matched together in a couple such that no element x of couple a prefers an element y of different couple b that also prefers x (i.e. the stability condition of the pairing). If such an (x, y) exists in the match, then it is defined as *blocking*; a match is stable if no blocking pairs exist.

The problem was first studied by Gale and Shapley [9]. They showed that there always exists at least a stable matching in any instance and they also proposed a $\mathcal{O}(n^2)$-time algorithm to find one, i.e. the so-called *Gale-Shapley* (GS) algorithm. An extended version of the GS algorithm, i.e. the EGS algorithm [13], avoids some unnecessary steps by deleting from the preference lists certain (man, woman) pairs that cannot belong to a stable matching. Notice that, in the man-oriented version of the EGS algorithm, each man has the best partner (according to his ranking) that he could obtain, whilst each woman has the worst partner that she can accept. Similar considerations hold for the woman-oriented version of EGS, where men have the worst possible partner.

For this reason, the classical problem has been extended [9] in order to find a SM under a more equitable measure of optimality, thus obtaining an *Optimal SM* problem [12,15,16,19]. For example, in [15] the authors maximize the total satisfaction in a SM by simply summing together the preferences of both men, $p_M(m_i, w_j)$, and women, $p_W(m_i, w_i)$, in the SM given by $MT = \{(m_i, w_j), \ldots, (m_k, w_z)\}$. This sum needs to be minimized since $p_M(m_i, w_j)$ represents the rank of w_j in m_i's list of preferences, where a low rank position stands for a higher preference, i.e. 1 belongs to the most preferred partner; similar considerations hold for the preferences of women, $p_W(m_i, w_j)$, which represents the rank of m_i in w_i's list of preferences. Therefore, we need to minimize this *egalitarian cost* [15]:

$$min \left(\sum_{(m_i, w_j) \in MT} p_M(m_i, w_j) + \sum_{(m_i, w_j) \in MT} p_W(m_i, w_j) \right) \qquad (1)$$

This optimization problem was originally posed by Knuth [15]. Other optimization criteria are represented by minimizing the *regret cost* [12] as in (2):

$$\min_{(m_i, w_j) \in MT} \max \max\{p_M(m_i, w_j), p_W(m_i, w_j)\} \qquad (2)$$

or by minimizing the *sex-equalness cost* [17] as in (3):

$$min \left| \sum_{(m_i, w_j) \in MT} p_M(m_i, w_i) - \sum_{(m_i, w_j) \in MT} p_W(m_i, w_j) \right| \qquad (3)$$

Even though the number of stable matchings for one instance grows exponentially in general [16], (1) and (2) have been already solved in polynomial time using ad-hoc algorithms such as [15] and [12], respectively, by exploiting a lattice structure that condenses the information about all matchings. On the contrary, (3) is an NP-hard problem for which only approximation algorithms have been given [17].

In the following, we consider preference as a more general weight, taken from a semiring, instead of a position in the preference's list of an individual; thus, we suppose to have *weighted preference lists* [15]. A different but compatible, with respect to OSM, variant of the SM problem allows incomplete preference's lists, i.e. the SM with incomplete lists (*SMI*), if a person can exclude a partner whom she/he does not want to be matched with [16]. Another extension is represented by preference lists that allow ties, i.e. in which it is possible to express the same preference for more than one possible partner: the problem is usually named as SM with ties, i.e. *SMT* [16]. In this case, three stability notions can be proposed [16]:

– Given any two couples (m_i, w_j) and (m_k, w_z), in a *super* stable match a pair (m_i, w_z) is blocking iff $p_M(m_i, w_z) \geq p_M(m_i, w_i) \wedge p_W(m_i, w_z) \geq p_W(m_k, w_z)$;
– In a *strongly* stable match a pair (m_i, w_z) is blocking iff $p_M(m_i, w_z) > p_M(m_i, w_i) \wedge p_W(m_i, w_z) \geq p_W(m_k, w_z)$ or $p_M(m_i, w_z) \geq p_M(m_i, w_i) \wedge p_W(m_i, w_z) > p_W(m_k, w_z)$; and
– In a *weakly* stable match a pair (m_i, w_z) is blocking iff $p_M(m_i, w_z) > p_M(m_i, w_i) \wedge p_W(m_i, w_z) > p_W(m_k, w_z)$.

Hence, if a match is super stable then it is strongly stable, and if it is strongly stable then it is weakly stable [16]. Allowing ties in preferences means that objectives (1), (2) and (3) above become hard even to approximate [16]. By joining together these two extensions, we obtain the *SMTI* problem: *SM with Ties and Incomplete lists* [16].

The preferences of men and women can be represented with two matrices M and W, respectively, as in Figure 2. A subset of these two matrices (for sake of simplicity) is represented in Figure 1 as a bipartite graph, where only the preferences of m_1, m_2, w_1 and w_2 are shown. For instance, the match $\{(m_1, w_2), (m_2, w_1)\}$ is not stable since (m_1, w_1) is a blocking pair: $p_M(m_1, w_1) < p_M(m_1, w_2) \wedge p_W(m_1, w_1) < p_W(m_2, w_1)$, i.e. $1 < 4 \wedge 1 < 4$ (here we use $<$ instead of $>$ because lower values are preferred).

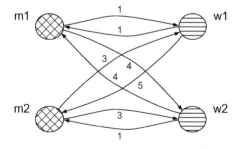

Fig. 1. An OSM problem represented as a bipartite graph

4   Representing the OSM Problem with Soft Constraints

In order to define an encoding of an OSM instance I as a SCSP instance P (see Section 2), we introduce the set V of variables: $m_1, m_2, \ldots, m_n$ corresponding to men, and $w_1, w_2, \ldots, w_n$ corresponding to women. The domain D of m_i or w_j is $[1, n]$. For each i, j $(1 \leq i, j \leq n)$, then $\eta : V \to D$ (as defined in Section 2) denotes the value of variable m_i and w_j respectively, i.e., the partner associated with the match. For example, $\eta(m_1) = 3$ means that m_1 is matched with w_3.

We need three different set of soft constraints to describe an OSM problem, according to each of the relationships we need to represent:

1. *Preference* constraints. These unary constraints represent the preferences of men and women: for each of the values in the variable domain, i.e. for each possible partner, they associate the relative preference. For example, $c_{m_i}(m_i = j) = a$ represents the fact that the man m_i has a degree of preference value a for the woman w_j (when the variable m_i is instantiated to j); on the other hand, $c_{w_j}(w_j = i) = b$ means that the same woman (w_j) has a preference for the same man (m_i) equal to b; a and b are elements of the chosen semiring set. We need $2n$ unary constraints: one for each man and woman.

2. *Marriage* constraints. This set constrains the marriage relationships: if m_i is married with w_j (i.e. $\eta(m_i) = j$), then w_j must be married with m_i (i.e. $\eta(w_j) = i$). Formally, it can be defined by $c_m(m_i, w_j) = \mathbf{0}$ if $\eta(m_i) = h \wedge \eta(w_j) = k \wedge (h \neq j \vee k \neq i)$. We need n^2 marriage constraints, one for each possible man-woman couple.

3. *Stability* constraints. This set of 4-ary constraints avoids the presence of blocking couples in the set of matches: $c_s(m_i, m_k, w_j, w_z) = \mathbf{0}$ if m_i and w_j are married (i.e. $\eta(m_i) = j$ and $\eta(w_j) = i$) and if there exists a different matched couple (m_k, w_z) (i.e. $k \neq i, z \neq j$ and $\eta(m_k) = z$ and $\eta(w_z) = k$) such that $c_{m_i}(m_i = j) <_S c_{m_i}(m_i = z) \wedge c_{w_z}(w_z = k) <_S c_{w_z}(w_z = i)$, where S represents the chosen semiring (see Section 2). In previous stability constraint definition we use $<_S$ because we are looking for a weakly stable marriage (see Section 3). For super and strong stabilities (see Section 3) we should instead define the stability constraints by using $\leq_S$. Therefore, we need n^4 stability constraints of this kind.

Given this encoding, the set of consistent solutions of P is equivalent to the set of solutions of I (i.e. an OSM problem instance). Therefore, unsatisfying the marriage or stability constraints makes P inconsistent (see Section 2). By using this formalization it is now possible to easily maximize the global satisfaction of all the couples, and thus finding a solution for the OSM problem. In practice it is possible to obtain the best possible solution of the SCSP problem being considered by exploiting the properties of the chosen semiring operators, i.e. $+$ and $\times$. For example, we could consider the preference as a cost, and the cost of the complete match could be obtained by summing together the costs of all the found (non-blocking) pairs. In this case, and if we want to minimize the cost of the n marriages, we can use the *Weighted* semiring [1,4], i.e. $\langle \mathbb{R}^+, min, \hat{+}, +\infty, 0 \rangle$

($\hat{+}$ is the arithmetic sum). Therefore, what we solve is exactly Objective (1) in Section 3.

Otherwise, we could use the *Fuzzy* semiring $\langle [0,1], max, min, 0, 1 \rangle$ [1,4] to maximize the "happiness" of the least sympathetic couple overall: the fuzzy values in the interval $[0,1]$ represent an "happiness degree" of the marriage relationships and are aggregated with min, but preferred with max. Again, what we solve with this semiring is exactly Objective (2) in Section 3, if we consider the ordering of the preferences as inverted (i.e. a high preference is better than a lower one); this is the reason why we use max − min instead of min − max.

Finally, as an example on the expressiveness of our framework, we can use the *Probabilistic* semiring $\langle [0,1], max, \hat{\times}, 0, 1 \rangle$ [1,4] ($\hat{\times}$ is the arithmetic multiplication) in order to maximize the probability that the obtained couples will not split. It is also possible to maximize the "happiness" of a fixed man or woman by setting to **1** the other preferences.

Moreover, we can represent the SMI extension reported in Section 3 by simply declaring a preference constraint with value corresponding to **0**: $c_{m_i}(m_i = j) = \mathbf{0}$ if m_i has not expressed a preference for w_j. Further on, by having the same value in the same preference list, i.e. $c_{m_i}(m_i = j) = a$ and $c_{m_i}(m_i = z) = a$, we can represent the SMT problem defined in Section 3. In Section 4.1 we consider and solve the most general problem among those presented in Section 3, i.e. the *Optimal SMTI (OSMTI)*.

Notice that such semiring structures allows us to consider also the preferences of men and women being partially ordered (see Section 2), which is clearly more generic and expressive with respect to the total ordering of the classical problem: Bob could love/like Alice and Chandra more than Drew, but he could not relate the first two girls with each other.

4.1 Specifying and Instance of the OSM Problem

In this section we solve the soft constraint formalization of the OSMTI problem given with preference, marriage and stability constraints. To achieve this goal, we represent and solve it as an ILP by using AMPL [8]. AMPL is a modeling language for mathematical programming with a very general and expressive syntax. It covers a variety of types and operations for the definition of indexing sets, as well as a range of logical expressions. The solution can be obtained with different solvers which can interface to AMPL; for our example we use CPLEX[1]. The soft constraints can be represented with AMPL statements. The obtained SCSP can be clearly solved also with other techniques as *branch-and-bound* [20], or branch-and-bound and *Symmetry Breaking via Dominance Detection* (SBDD) [5]; however, the ILP solver represents a completely new approach with respect to SCSP, and provides a bridge between the two fields.

We consider an instantiation of the (1) problem in Section 3, and therefore the adopted semiring is $\langle \mathbb{R}^+, min, \hat{+}, +\infty, 0 \rangle$, even if, as said before, we can also solve other criteria by changing the semiring. The two matrices M and W in

---

[1] http://www.ilog.com/products/cplex/

```
set MEN := m1 m2 m3 m4 m5 m6 ;
set WOMEN := w1 w2 w3 w4 w5 w6 ;

param M:                              param W:
      w1 w2 w3 w4 w5 w6 :=                  w1 w2 w3 w4 w5 w6 :=
m1    1  4  Inf 5  5  3           m1        1  4  6  2  4  2
m2    3  4  6  1  5  2            m2        5  1  4  5  2  6
m3    1  Inf 4  2  3  5           m3        4  5  2  2  Inf 3
m4    6  1  3  4  2  1            m4        4  2  1  4  5  5
m5    3  1  2  4  5  6            m5        2  6  5  Inf 6  1
m6    3  3  1  6  5  4 ;          m6        3  Inf 3  6  3  4 ;
```

Fig. 2. The data file of our example in AMPL: the sets of MEN and $WOMEN$ and their respective preference lists (M and W)

Figure 2 respectively represent the preference values of $n = 6$ men ($MEN = \{m_1, m_2, m_3, m_4, m_5, m_6\}$) and $n = 6$ women ($WOMEN = \{w_1, w_2, w_3, w_4, w_5, w_6\}$) taken from the Weighted semiring set. Notice that both M and W are displayed Figure 2 with men on rows and women on columns, in order to improve the readability when comparing the two matrices. The lists of preferences of men are represented by the rows of M, and the preferences of women are instead the columns of W.

Since we want to deal with incomplete lists, the preference value corresponds to the bottom element of the semiring (in Weighted semiring, it is ∞) if that preference has not been expressed; Inf in Figure 2 is a shortcut for a very large value that we can consider as the infinite value (e.g. 10000). For example, in Figure 2 $M[m_1, w_3] = \infty$ means that m_1 has no preference for w_3. Moreover, we can deal with ties at the same time, e.g. $M[m_4, w_2] = M[m_4, w_6] = 1$ in Figure 2.

Notice that this problem could have no solution in general due to the fact that the preference lists are incomplete and we want to find a perfect match (n pairs). Moreover, since we have ties and we require a weakly stable matching, the problem is NP-hard [16].

4.2 A Formalization as an Integer Linear Program

With AMPL we need to create two files storing the data of the problem (Figure 2) and its model (Figure 3). The *Marriage* variable in Figure 3 corresponds to the couples representing the best stable marriage, while the *EgalitarianCost* is exactly computed as for Objective (1) in Section 3 and the goal is to minimize it. Notice that by changing the mathematical operators of the *OBJECTIVE* in Figure 3, it is possible to solve also Objectives (2) and (3) of Section 3. The *MenMarriages* and *WomenMarriages* constraints state that each man and each woman must have a partner, respectively, that is we require a perfect match. At last, the *Stability* constraint prevents blocking pairs.

```
option solver cplex;

### PARAMETERS ###
set MEN;
set WOMEN;
param M {i in MEN, j in WOMEN};
param W {k in MEN, z in WOMEN};

### VARIABLES ###
var Marriage {i in MEN, j in WOMEN} binary;

### OBJECTIVE ###
minimize EgalitarianCost:  sum {i in MEN, j in WOMEN}
    (( Marriage[i,j] * M[i,j] ) +
    ( Marriage[i,j] * W[i,j] )) ;

### CONSTRAINTS ###
subject to MenMarriages {i in MEN}:
    sum {j in WOMEN} Marriage[i,j] = 1 ;
subject to WomenMarriages  {j in WOMEN}:
    sum {i in MEN} Marriage[i,j] = 1 ;
subject to Stability {i in MEN, k in MEN, j in WOMEN, z in WOMEN:
    ( M[i,z] < M[i,j] ) and
    ( W[i,z] < W[k,z] )}:
    Marriage[i,j] + Marriage[k,z] <= 1;
```

Fig. 3. The file storing the model for our example in AMPL

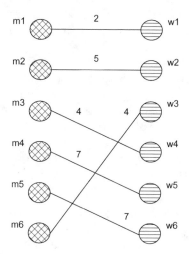

Fig. 4. The optimal stable match that can be obtained from the AMPL program in Figure 2 and Figure 3

The three marriages that can be obtained with this formalization are respectively $SM_1 = \{(m_1, w_1), (m_2, w_2), (m_3, w_4), (m_4, w_6), (m_5, w_5), (m_6, w_3)\}$, $SM_2 = \{(m_1, w_1), (m_2, w_2), (m_3, w_4), (m_4, w_3), (m_5, w_6), (m_6, w_5)\}$ and, at last, $SM_3 = \{(m_1, w_1), (m_2, w_2), (m_3, w_4), (m_4, w_5), (m_5, w_6), (m_6, w_3)\}$. The egalitarian costs for these three matches are respectively $ec(SM_1) = 32$, $ec(SM_2) = 30$ and $ec(SM_3) = 29$, which is also the result of the program in Figure 3 since it corresponds to the lowest possible cost. The SM_3 solution is also represented in Figure 4 as a bipartite graph, where the man/woman preferences within the same couple are added on the same edge, i.e. the cost of the edge (m_1, w_1) is $M[m_1, w_1] + W[m_1, w_2] = 2$ (the values in the matrices of Figure 2).

5   Multi-Agent Systems and the Stable Marriage of Coalitions

Cooperating groups, referred to as coalitions, have been thoroughly investigated in artificial intelligence and game theory and have proved to be useful in both real-world economic scenarios and multi-agent systems [14]. Coalitions, in general, are task-directed and short-lived, but last longer than team organization [14] (for example) and in some cases they have a long lifetime once created [11]. Given the population of entities E, the problem of coalition formation consists in selecting the appropriate partition of E, $P = \{C_1, \ldots, C_n\}$ ($|P| = |A|$ if each entity forms a coalition on its own), such that $\forall C_i \in P$, $C_i \subseteq E$ and $C_i \cap C_j = \emptyset$, if $i \neq j$. P maximizes the utility (utility against costs) that each coalition can achieve in the environment. Therefore, agents group together because utility can be gained by working in groups, but this growth is somewhat limited by the costs associated with forming and maintaining such a structure.

Cooperation involves a degree of risk arising from the uncertainties of interacting with autonomous self-interested agents. Trust [18] describes a node's belief in another node's capabilities, honesty and reliability based on its own direct experiences. Therefore trust metrics have been already adopted to perceive this risk, by estimating how likely other agents are to fulfill their cooperative commitments [7,11]. Since trust is usually associated with a specific scope [18], we suppose that this scope concerns the task that the coalition must face after its formation; for example, in electronic marketplaces the agents in the same coalition agree with a specific discount for each transaction executed [7,21]. Clearly, an entity can also trust itself in achieving the task, and can form a singleton coalition.

5.1   Defining the Stable Marriage for Coalitions

In an individual-oriented approach an agent prefers to be in the same coalition with the agent with whom it has the best relationship [7]. In socially-oriented classification the agent instead prefers the coalition in which it has most summative trust [7]. In this Section we would like to rephrase the classical notion of stability in SM problems (presented in Section 3) as coalition formation criteria. Moreover, instead of a preference (as in Section 3), we need to consider a trust relationship between two entities, which, inherently expresses a preference in some sense. To do so, in Definition 1 we formalize how to compute the trustworthiness of a whole coalition:

Definition 1 (Trustworthiness of a Coalition). *Given a coalition C of agents defined by the set $\{x_1, \ldots, x_n\}$ and a trust function t defined on ordered pairs (i.e. $t(x_i, x_y)$ is the trust score that x_i associates with x_j), the trustworthiness of C (i.e. $T(C)$) is defined as the composition (i.e. $\circ$) of the 1-to-1 trust relationships, i.e. $\forall x_i, x_j \in C. \circ t(x_i, x_j)$ (notice that i can be equal to j, modeling an agent's trust in itself).*

The $\circ$ function has already been defined in [6]; it models the composition of the 1-to-1 trust relationships. It can be used to consider also subjective ratings [18] (i.e. personal points of view on the composition), even if in this paper we will consider objective ratings [18] in order to easily represent and compute trust with a mathematical operator. For instance, some practical instantiations of the $\circ$ function can be the *arithmetic mean* or the *max* operator: $\forall x_i, x_j \in C. avg\, t(x_i, x_j)$ or $\forall x_i, x_j \in C. \max t(x_i, x_j)$. Notice that the $\circ$ operation is not only a plain "addition" of the single trust values, but it must also take into account also the "added value" (or "subtracted value") derived from the combination effect.

As proposed in Section 4 for the classical problem, by changing the semiring structure we can represent different trust metrics [6,22]. Therefore, the optimization of the set of coalitions can follow different principles, as, for example, minimizing a general cost of the aggregation or maximizing the "consistency" evaluation of the included entities, i.e. how much their interests are alike. In order to extend the stability condition of the classical problem, blocking coalitions are defined in Definition 2:

Definition 2 (Blocking Coalitions). *Two coalitions C_u and C_v are defined as blocking if, an individual $x_k \in C_v$ exists such that, $\forall x_i \in C_u, x_j \in C_v$ with $j \neq k$, $\circ_{x_i \in C_u} t(x_k, x_i) > \circ_{x_j \in C_v} t(x_k, x_j)$ and $T(C_u \cup x_k) > T(C_u)$ at the same time.*

Clearly, a set $\{C_1, C_2, \ldots, C_n\}$ of coalitions is *stable* if no blocking coalitions exist in the partitioning of the agents. An example of two blocking coalitions is

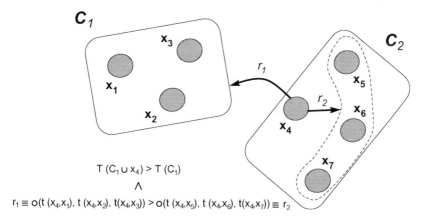

Fig. 5. A graphical intuition of two blocking coalitions

sketched in Figure 5: if x_4 prefers the coalition C_1 (relationship r_1 in Figure 5) to the elements in its coalitions C_2 (r_2 in Figure 5), i.e. $\circ(t(x_4, x_1), t(x_4, x_2), t(x_4, x_3)) > \circ(t(x_4, x_5), t(x_4, x_6), t(x_4, x_7))$, and C_1 increases its trust value by having x_4 inside itself, i.e. $T(C_1 \cup x_4) > T(C_1)$, then C_1 and C_2 are two blocking coalitions and the partitioning $\{C_1, C_2\}$ is not stable and thus, it is not a feasible solution of our problem.

We therefore require the stability condition to be satisfied, but at the same time want to optimize the trustworthiness of the partition given by aggregating together all the trustworthiness scores of the obtained coalitions.

5.2 A Formalization of the Problem

As accomplished in Section 4 for the classical problem, in this Section we define the soft constraints needed to represent the coalition-extension problem. As an example, we adopt the *Fuzzy* semiring $\langle [0, 1], max, min, 0, 1 \rangle$ in order to maximize the minimum trustworthiness of all obtained coalitions (as proposed also in [2,6]). The following definition takes the general $\circ$ operator (presented in Section 5) as one of its parameters: it can be considered in some sense as a "lower level" operator with respect to the other two semiring operators (i.e. $+$ and $\times$).

The variables V of this problem are represented by the maximum number of possible coalitions: $\{co_1, co_2, \ldots, co_n\}$ if we have to partition a set $\{x_1, x_2, \ldots, x_n\}$ of n elements. The domain D for each of the variables is the powerset of the element identifiers, i.e. $\mathcal{P}\{1, 2, \ldots, n\}$; for instance, if $\eta(co_1) = \{1, 3, 5\}$ it means the the coalition co_1 groups the elements x_1, x_2, x_5 together Clearly, $\eta(co_i) = \emptyset$ if the framework finds less than n coalitions.

1. *Trust* constraints. As an example from this class of constraint, the soft constraint $c_t(co_i = \{1, 3, 5\}) = a$ quantifies the trustworthiness of the coalition formed by $\{x_1, x_3, x_5\}$ into the semiring value represented by a. According to Definition 1, this value is obtained by using the $\circ$ operator and composing all the 1-to-1 trust relationships within the coalition. In this way we can find the best set of coalitions according to the semiring operators. This kind of constraint resembles the preference constraints given in Section 4.

2. *Partition* constraints. This set of constraints is similar to the *Marriage* constraints proposed in Section 4. It is used to enforce that an element belongs only to one single coalition. For this goal we can use a binary crisp constraint between any two coalition, as $c_p(co_i, co_j) = 0$ if $\eta(co_i) \cap \eta(co_j) \neq \emptyset$, and $c_p(co_i, co_j) = 1$ otherwise (with $i \neq j$). Moreover, we need to add one more crisp constraint in order to enforce that all the elements are assigned to at least one coalition: $c_p(co_1, co_2, \ldots, co_n) = 0$ if $|\eta(co_1) \cup \eta(co_2) \cup \cdots \cup \eta(co_n)| \neq n$, and $c_p(co_1, co_2, \ldots, co_n) = 1$ if $|\eta(co_1) \cup \eta(co_2) \cup \cdots \cup \eta(co_n)| = n$.

3. *Stability* constraints. These crisp constraints model the stability condition extended to coalitions, as proposed in Definition 2. We have several ternary constraints for this goal: $c_s(co_v, co_u, x_k) = 0$ if $k \in \eta(co_v)$ (i.e. x_k belongs to the co_v coalition), $\circ_{i \in \eta(co_u)} t(x_k, x_i) > \circ_{j \in \eta(co_v)} t(x_k, x_j)$ and $c_t(\eta(co_u) \cup k) > c_t(co_u)$. Otherwise, $c_s(co_v, co_u, x_k) = 1$.

6   Conclusions

In this paper we have presented a general soft constraint-based framework to represent and solve the Optimal Stable Marriage (OSM) problem [15] and its variants with incomplete preference lists or ties amongst preferences. The optimization criterion depends on the chosen semiring (e.g. *Weighted* or *Fuzzy*) which can be used to solved problems already proposed in literature, such as minimizing the egalitarian cost (see Section 3 and Section 4). Therefore, it is possible to solve all these different optimization problems with the same general framework, and we do not need an ad-hoc algorithm for each distinct case (e.g. [12,15,17]). One of the aims of this paper was to relate the OSM and soft constraint satisfaction as done also for the classical SM and classic constraint satisfaction [10,23]. Since many variants of the OSM problem are NP-hard [16], representing and solving the problem as a SCSP can be a valuable strategy [10]. Integer linear programming, the tool adopted to find a solution for the related soft constraint problem, was applied here to this kind of problems for the first time.

Moreover, we have extended the OSM problem to achieve stable coalitions of agents/individuals by using trust metrics as a way to express preferences. Thus, we extend the stability conditions from agent-to-agent to agent-to-coalition (of agents); in this case the marriage is between an agent and a group of agents. What we obtain is a partition of the set of agents into trusted coalitions, such that no agent or coalition is interested in breaking the current relationships and consequently changing the partition. As future work, we would like to also use ILP to solve the problem extension to coalition formation, which has been modeled in Section 5.2. Moreover, we would like to compare the performance of the ILP framework with other classical SCSP solvers based on branch-and-bound procedures [5,20].

It would be interesting try to extend the results of this paper by modeling the formation and the consequent behaviour of the other organizational paradigms presented in [14], e.g. Holoarchies, Federations or Teams. To do so, we need to represent the different grouping relationships among the entities with soft constraints. We would like also to further explore the strong links between OSM and Games Theory, for example by developing even more sophisticated notions of stability.

Acknowledgements

This work was supported, in part, by Science Foundation Ireland (Grant Number 05/IN/I886). This work is based on the results of a research visit that Francesco Santini spent at the Cork Constraint Computation Centre.

References

1. Bistarelli, S.: Semirings for Soft Constraint Solving and Programming. LNCS, vol. 2962. Springer, Heidelberg (2004)
2. Bistarelli, S., Martinelli, F., Santini, F.: A semantic foundation for trust management languages with weights: An application to the RT family. In: Rong, C., Jaatun, M.G., Sandnes, F.E., Yang, L.T., Ma, J. (eds.) ATC 2008. LNCS, vol. 5060, pp. 481–495. Springer, Heidelberg (2008)

3. Bistarelli, S., Montanari, U., Rossi, F.: Soft concurrent constraint programming. ACM Trans. Comput. Logic 7(3), 563–589 (2006)
4. Bistarelli, S., Montanari, U., Rossi, F.: Semiring-based Constraint Solving and Optimization. Journal of the ACM 44(2), 201–236 (1997)
5. Bistarelli, S., O'Sullivan, B.: Combining branch & bound and SBDD to solve soft CSPs. In: Proc. of CP 2004 Fourth International Workshop on Symmetry and Constraint Satisfaction Problems (SymCon 2004) (2004)
6. Bistarelli, S., Santini, F.: Propagating multitrust within trust networks. In: SAC 2008: Proceedings of the 2008 ACM symposium on Applied computing, pp. 1990–1994. ACM, New York (2008)
7. Breban, S., Vassileva, J.: A coalition formation mechanism based on inter-agent trust relationships. In: AAMAS 2002: Proceedings of the first international joint conference on Autonomous agents and multiagent systems, pp. 306–307. ACM, New York (2002)
8. Fourer, R., Gay, D.M., Kernighan, B.W.: AMPL: A Modeling Language for Mathematical Programming. Brooks/Cole Publishing Company, Monterey (2002)
9. Gale, D., Shapley, L.S.: College admissions and the stability of marriage. The American Mathematical Monthly 69(1), 9–15 (1962)
10. Gent, I.P., Irving, R.W., Manlove, D., Prosser, P., Smith, B.M.: A constraint programming approach to the stable marriage problem. In: Conference on Principles and Practice of Constraint Programming, pp. 225–239. Springer, London (2001)
11. Griffiths, N., Luck, M.: Coalition formation through motivation and trust. In: AAMAS 2003: Proceedings of the second international joint conference on Autonomous agents and multiagent systems, pp. 17–24. ACM, New York (2003)
12. Gusfield, D.: Three fast algorithms for four problems in stable marriage. SIAM J. Comput. 16(1), 111–128 (1987)
13. Gusfield, D., Irving, R.W.: The stable marriage problem: structure and algorithms. MIT Press, Cambridge (1989)
14. Horling, B., Lesser, V.: A survey of multi-agent organizational paradigms. Knowl. Eng. Rev. 19(4), 281–316 (2004)
15. Irving, R.W., Leather, P., Gusfield, D.: An efficient algorithm for the "optimal" stable marriage. J. ACM 34(3), 532–543 (1987)
16. Iwama, K., Miyazaki, S.: A survey of the stable marriage problem and its variants. In: International Conference on Informatics Education and Research for Knowledge-Circulating Society (icks 2008), pp. 131–136. IEEE Computer Society Press, Los Alamitos (2008)
17. Iwama, K., Miyazaki, S., Yanagisawa, H.: Approximation algorithms for the sex-equal stable marriage problem. In: Dehne, F., Sack, J.-R., Zeh, N. (eds.) WADS 2007. LNCS, vol. 4619, pp. 201–213. Springer, Heidelberg (2007)
18. Jøsang, A., Ismail, R., Boyd, C.: A survey of trust and reputation systems for online service provision. Decis. Support Syst. 43(2), 618–644 (2007)
19. Knuth, D.E.: Stable marriage and its relation to other combinatorial problems: An introduction to the mathematical analysis of algorithms. American Mathematical Society, Providence (1997)
20. Leenen, L., Ghose, A.: Branch and bound algorithms to solve semiring constraint satisfaction problems. In: Ho, T.-B., Zhou, Z.-H. (eds.) PRICAI 2008. LNCS, vol. 5351, pp. 991–997. Springer, Heidelberg (2008)

21. Lerman, K., Shehory, O.: Coalition formation for large-scale electronic markets. In: ICMAS, pp. 167–174. IEEE Computer Society, Los Alamitos (2000)
22. Theodorakopoulos, G., Baras, J.S.: Trust evaluation in ad-hoc networks. In: WiSe 2004: Proceedings of the 3rd ACM workshop on Wireless security, pp. 1–10. ACM, New York (2004)
23. Unsworth, C., Prosser, P.: Specialised constraints for stable matching problems. In: van Beek, P. (ed.) CP 2005. LNCS, vol. 3709, p. 869. Springer, Heidelberg (2005)

Solving CSPs with Naming Games

Stefano Bistarelli[1,2,3] and Giorgio Gosti[3]

[1] Dipartimento di Scienze, Università "G. d'Annunzio" di Chieti-Pescara, Italy
bista@sci.unich.it
[2] Institute of Informatics and Telematics (IIT-CNR) Pisa, Italy
{stefano.bistarelli}@iit.cnr.it
[3] Dipartimento di Matematica e Informatica,
Università degli Studi di Perugia
{giorgio.gosti}@dipmat.unipg.it

Abstract. Constraint solving problems (CSPs) represent a formalization of an important class of problems in computer science. We propose here a solving methodology based on the naming games. The naming game was introduced to represent N agents that have to bootstrap an agreement on a name to give to an object. The agents do not have a hierarchy and use a minimal protocol. Still they converge to a consistent state by using a distributed strategy. For this reason the naming game can be used to untangle distributed constraint solving problems (DCSPs). Moreover it represents a good starting point for a systematic study of DCSP methods, which can be seen as further improvement of this approach.

1 Introduction

The goal of this research is to generalize the naming game model in order to define a distributed method to solve CSPs. In the study of this method we want to fully exploit the power of distributed calculation, by letting the CSP solution emerge, rather than being the conclusion to a sequence of statements.

In DCSP protocols we design a distributed architecture of processors, or more generally a group of agents, to solve a CSP instantiation. In this framework we see the problem as a dynamic system and we set the stable states of the system as one of the possible solutions to our CSP. To do this we design each agent in order to move towards a stable local state. The system is called "self-stabilizing" whenever the global stable state is obtained starting from local stable state [2]. When the system finds the stable state the CSP instantiation is solved. A protocol designed in this way is resistant to damage and external threats because it can react to changes in the problem.

In Section 2 we illustrate the naming game formalism and we make some comparisons with the distributed CSP (DCSP) architecture. Then we describe the language model that is common to the two formalizations and introduce an interaction scheme to show the common framework. At last we state the definition of Self-stabilizing system [2].

A. Oddi, F. Fages, and F. Rossi (Eds.): CSCLP 2008, LNAI 5655, pp. 16–32, 2009.

In Section 3 we explicitly describe our generalization and formalize the protocol that our algorithm will use and test it on different CSPs. Moreover, for particular CSPs instantiations we analytically describe the multi-agent algorithm evolution that makes the system converge to the solution.

2   Background

2.1   The Distributed Constraint Satisfaction Problem (DCSP)

Each constraint satisfaction problem (CSP) is defined by three sets $\langle X, D, C \rangle$: X is a set of N variables $x_1, x_2, \ldots x_N$, D is the set of the definition domains $D_1, D_2, \ldots, D_N$ of the variables in X, and C is a set of constraints on the values of these variables. Each variable X_i is defined in its variable domain D_i with i taking integer values from 1 to N. Each constraint is defined as a predicate on the values of a sub-set of our variables $P_k(x_{k1}, x_{k2}, \ldots x_{kM})$. The indices $k1, k2, \ldots kM$ with $M < N$, are a sequence of strictly increasing integers from 1 to M and denote the sub-set of our variables $x_{k1}, x_{k2}, \ldots x_{kM}$. The Cartesian product of these variable domains $D_{k1} \times D_{k2} \times \ldots \times D_{kM}$ is the domain of our predicate. The predicate P_k is true only for a fixed subset T of its domain. When the values assigned to the variables of the predicate P_k are in this subset T, the predicate is true and we say that the constraint is satisfied. A CSP solution is a particular tuple $\overline{X}$ of the $x_1, x_2, \ldots x_N$ variable assignments that satisfy all the constraints C.

In the DCSP [5], the variables of the CSP are distributed among the agents. These agents are able to communicate between themselves and know all the constraint predicates that are relevant to their own variables. The agents through interaction find the appropriate values to assign to the variables and solve the CSP.

2.2   Introduction to Naming Games

The naming games [7,9,10] describe a set of problems in which a number N of agents bootstrap a commonly agreed name for an object. Each naming game is defined by an *interaction protocol*. An important aspect of the naming game is the hierarchy-free agent architecture. The naming task is achieved through a sequence of interactions in which two agents are randomly extracted at each turn to perform the role of the speaker and the listener (or hearer as used in [7,9]). The speaker declares its name suggestion for the object. The listener receives the word and computes the communication outcome. The communication outcome is determined by the *interaction protocol*, in general it depends on the previous interactions of the listener and if it agrees or disagrees with the name assignment. The listener will express the communication outcome, which determines the agents update at the end of each turn. The agents in this way change their internal state at each turn through interaction. DCSP and the naming game share a variety of common features [1], moreover we will show in Section 3 that the naming game can be seen as a particular DCSP.

2.3   The Communication Model

In this framework we define a general model that describes the communication procedures between agents both in naming games and in DCSPs. The communication model consists of N agents (also called processors) arranged in a network.

The systems that we consider are self-stabilizing and evolve through interactions in a stable state. We will use a central scheduler that at each turn randomly extracts the agents that will be interacting.

The network links connect agents that can communicate with each other; this network can be viewed as a *communication graph*. Each link can be seen as a register r_{ij} on which the speaker i writes the variable assignment or word it wants to communicate, and the listener j can read this assignment. We assume that the two communication registers $r_{ij} \neq r_{ji}$ are different and that each communication register can have more then one field. We also define a general communication register in which only the speaker i can write and can be read by all the neighboring listeners. This is the convention which we will use since in our algorithm at each interaction the speaker communicates the same variable assignment (word) to all the neighbors. For each link of the *communication graph* r_{ij} we allocate a register f_{ij} so the listener can give feedback on the communication outcome using a predetermined signaling system.

The interaction scheme can be represented in three steps:

1. *Broadcast.* The speakers broadcast information related to the proposed assignment for the variable;
2. *Feedback.* The listeners feedback the interaction outcome expressing some information on the speaker assignment by using a standardized signal system;
3. *Update.* The speakers and the listeners update their state regarding the overall interaction outcome.

In this scheme we see that at each turn the agents update their state. The update reflects the interaction they have experienced. In this way the agent communication makes the system self-stabilizing. We have presented the general interaction scheme, wherein each naming game and DCSP algorithm has its own characterizing protocol.

2.4   Self-stabilizing Algorithms

A self-stabilizing protocol [2] has some important propierties. First, the global stable states are the wanted solutions to our problem. Second, the system configurations are divided into two classes: legal associated to solutions and illegal associated to non-solutions. We may define the protocol as self-stabilizing if in any infinite execution the system finds a legal system configuration that is a global equilibrium state. Moreover, we want the system to converge from any initial state. These properties make the system fault tolerant and able to adapt its solutions to changes in the environment.

To make a self-stabilizing algorithm we program the agents of our distributed system to interact with the neighbors. The agents through these interactions update their state trying to find a stable state in their neighborhood. Since the algorithm is distributed many legal configurations of the agents' states and its neighbors' states start arising sparsely. Not all of these configurations are mutually compatible and so form incompatible legal domains. The self-stabilizing algorithm must find a way to make the global legal state emerge from the competition between this domains. Dijkstra [2] and Collin [6] suggest that an algorithm designed in this way can not always converge and a special agent is needed to break the system symmetry. In this paper we will show a different strategy based on the concept of random behavior and probabilistic transition function that we will discuss in the next sections.

3   Generalization of the Naming Game to Solve DCSP

In the naming game, the agents want to agree on the name given to an object. This can be represented as a DCSP, where the name proposed by each agent is the assignment of the CSPs variable controlled by the agent, and where an equality constraint connects all the variables. On the other hand, we can generalize the naming game to solve DCSPs.

We attribute an agent to each variable of the CSP as in [5]. Each agent $i = 1, 2, \ldots N$, names its own variable x_i in respect to the *variable domain* D_i. We restrict the constraints to binary relation C_{ij} between variable x_i and x_j. This relation can be an equality (to represent the naming game), an inequality, or any binary relation. If $x_i C_{ij} x_j$ is true, then the values of the variables x_i and x_j are consistent. We define two agents as neighbors if their variables are connected by a constraint.

The agents have a *list*, which is a continuously updated subset of the domain elements. The difference between the *list* and the domain is that the domain is the set of values introduced by the problem instance, and the *list* is the set of variable assignments that the agent subjectively forecasts to be in the global solution, on the basis of its past interactions. When the agent is a speaker, it will refer to this *list* to choose the value to broadcast and when it is a listener, it will use this *list* to evaluate the speaker broadcasted value.

At turn $t = 0$ the agents start an empty *list*, because they still do not have information about the other variable assignments. At each successive turn $t = 1, 2, \ldots$ an agent is randomly extracted by the central scheduler to cover the role of the speaker, and all its neighbors will be the listeners. The communication between the speaker s and a single listener l can be a *success*, a *failure*, or a *consistency failure*. Let d_s and d_l be respectively the speaker's and the listener's assignment. *Success* or *failure* is determined when the variable assignments satisfy or not the relation $d_s C_{sl} d_l$. *Consistency failure* occurs when the listener does not have any assignment in its *variable domain* that is consistent with the proposed speaker variable assignment.

At the end of the turn all the listeners communicate to the speaker the *success*, the *failure*, or the *consistency failure* of the communication.

If all the interaction sessions of the speakers with the neighboring listeners are successful, we will have a *success update*: the speaker eliminates all the assignments and keeps only the successful assignment; the listeners eliminate all the assignments that are not consistent to the successful assignment of the speaker. If there was one or more *consistency failure* the speaker eliminates its variable assignment from the variable domain, we call this a *consistency failure update*. If there was no *consistency failure* and just one or more *failures*, there will be a *failure update*: the listeners update their lists adding to the set of possible assignments the set of consistent assignments of the speaker utterance.

The interaction, at each turn t, is represented by this protocol:

1. *Broadcast.* If the speaker *list* is empty it extracts an element from its *variable domain* D_s, puts it in its *list* and communicates it to the neighboring listeners. Otherwise, if its *list* is not empty, it randomly draws an element from its *list* and communicates it to the listeners. We call the broadcast assignment d_s.
2. *Feedback.* Then the listeners calculate the consistant assignment subset K and the consistant domain subset K':
   - *Consistency evaluation.* Each listener uses the constraint defined by the edge, which connects it to the speaker, to find the consistent elements d_l to the element d_s received from the speaker. The elements d_l that it compares with d_s are the elements of its *list*. These consistent elements form the *consistent elements subset* K. We define $K = \{d_l \in list | d_s C_{sl} d_l\}$. If K is empty the listeners compare each element of its variable domain D_l with the element d_s, to find a *consistent domain subset* K'. We define $K' = \{d_l \in D_l | d_s C_{sl} d_l\}$.

   The *consistent elements subset* K and the *consistent domain subset* K' determine the following feedback:
   - *Success.* If the listener has a set of elements d_l consistent to d_s in its *list* (K is not empty), there is a *success*.
   - *Consistency failure.* If the listener does not have any consistent elements d_l to d_s in its *list* (K is empty), and if no element of the listener *variable domain* is consistent to d_s (K' is empty), there is a *consistency failure*.
   - *Failure.* If the listener does not have any d_l consistent elements to d_s in its *list* (K is empty), and if a non empty set of elements of the listener *variable domain* are consistent to d_s (K' is not empty), there is a *failure*.
3. *Update.* Then we determine the overall outcome of the speaker interaction on the basis of the neighbors' feedback:
   - *Success update.* This occurs when all interactions are successful. The speaker and the neighbors cancel all the elements in their *list* and update it in the following way: the speaker stores only the successful element d_s and the listener stores the consistent elements in K.
   - *Consistency failure update.* This occurs when there is at least one *consistency failure* interaction. The speaker must eliminate the element d_s from its *variable domain* (this can be seen as a step of local consistency pruning). The listeners do not change their state.
   - *Failure update.* This occurs in the remaining cases. The speaker does not update its list. The listeners update their *lists* by adding the set K' of all the elements consistent with d_s to the elements in the *list*.

We can see that in the cases where the constraint $x_i C_{ij} x_j$ is an equality, the subset of consistent elements to x_i is restricted to one assignment of x_j. For this assignment of the constraint $x_i C_{ij} x_j$ we obtain the naming game as previously described. Our contribution to the interaction protocol is to define K and K' since in the naming game the consistent listener assignment d_l to the speakers assignment d_s is one and only one ($d_l = d_s$). This is fundamental to solve general CSP instances. Moreover, in the naming game there is only one speaker and one listener at each turn. Under this hypothesis the agents were not always able to enforce local consistency (e.g.: graph coloring of a completely connected graph). Thus we had to extend the interaction to all the speaker neighbors and let all the neighbors be listeners.

At each successive turn the system evolves through the agents interactions in a global equilibrium state. In the equilibrium state all the agents have only one element for their variable and this element must satisfy the constraint $x_i C_{ij} x_j$ with the element chosen by the neighboring agents. We call this the state of *global consensus*. Once in this state the interactions are always successful. The probability to transit to a state different from the *global consensus* state is zero, for this reason the *global consensus* state is referred to as an absorbing state. We call the turn at which the system finds global consensus *convergence turn* t_{conv}.

Interaction Example. As an example we can think of the interaction between a speaker and its neighbors on a graph coloring instance. If the speakers *list* is empty its draws a random element from its domain and puts it the *list*. If the speakers *list* is not empty he draws a random element from the elements on this *list*. Let's say he picks the color red this will be its broadcast $d_s = $ 'red'. Each listener will use this value to compute K and K'. First the listener will determine which elements in its *list* are different from red: $K = \{d_h \in list | \text{'red'} \neq d_h\}$. If K is empty it will find the elements on its domain D_h which are different from red: $K' = \{d_h \in D_h | \text{'red'} \neq d_h\}$. Once the listener has calculated K and K' he can determine the feedback. If K is not empty the listener will feedback a success. If K' is not empty the listener will feedback a failure. If K and K' are empty the listener will feedback a consistency failure. At this point the speaker will use the feedback information to choose the update modality. If all the listeners feedback *success*, this means that they have colors different from red in their *list*. The speaker chooses to have a success update and this means that it deletes all the colors from its *list* and keeps only the element red. The listeners contrarily delete the red element in their *list* if there is one. If one or more listeners feedback a failure then the speaker will not change its *list*, but the listeners will add the elements in K' in their *list*. In this case, this means that the listener will have in its list all the colors different from red, plus red, if this color was already in the listener *list*.

Simple Algorithm Execution. The N-Queens Puzzle is the problem of placing N queens on a $N \times N$ chessboard without having them be mutually capturable. This means that there can not be two queens that share the same row, column, or diagonal. This is a well known problem and has been solved linearly

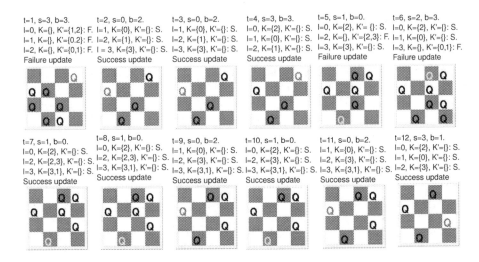

Fig. 1. Single algorithm run for the N-Queens Puzzle with $N = 4$

by specialized algorithms. Nevertheless it is considered a classical benchmark and we use it to show how our algorithm can solve different instances. To reduce the search space we assign the variables of a different column to each queen. We can do this because if there were more then one queen in a column, they would have been mutually capturable. In this way each of our agents will have as its domain the values of a distinct column and all the agents will be mutually connected by an edge in the graph representing all the constraints. In the example we show a N-Queens Puzzle with $N = 4$. Each agent (queen) is labeled after its column with numbers from zero to three from left to right. The rows are labeled from the bottom with numbers from zero to three. In the Figure 1 we show how the algorithm explores the solution space randomly and how it evolves at each turn t. We write the speaker s that is extracted at each turn and its broadcasted value b. Then we write the listeners l, their respective K, K', and the feedbacks. At the end we write the updates. The picture represents graphically the evolution of the agent's list at the end of each turn.

At turn $t = 1$ (see Fig.1) speaker $s = 3$ is randomly drawn. The *variable* controlled by this speaker is the position of the queen on the last column of the chessboard. The speaker has an empty *list*, hence it draws from its *variable domain* the element $d_s = 3$ which corresponds to the highest row of its column. The speaker add this new element to its *list*. Since all the agent are connected, all the agents apart from the speaker are listeners. Their *lists* are empty therefore K is empty. Thus they compute K' from the *variable domain*. The listeners feedback failure, thus the speaker replies with a failure update. The listeners add the elements of their respective K' to their *lists*. The picture in fig.1 shows the elements in the agents' *lists* at the end of the turn. At turn $t = 2$ speaker $s = 0$ is drawn and it broadcasts the element $d_s = 2$. All the listeners have a consistent element in their list, therefore, their Ks are not empty, and they feedback a success. The listeners delete their *lists* and add the elements in their K. At turn $t = 3$ the speaker $s = 0$ speaks again and broadcasts the same element $d_s = 2$. Therefore, the listener computes the same Ks of before, and feedback a success. Thus we have a success update but since the Ks are the same the system does not change. At turn $t = 4$ the speaker $s = 0$ is drawn and broadcast the same variable $d_s = 3$ that it had broadcasted at the first turn. Since all the elements in the listeners' *lists* are still consistent to this broadcast, the algorithm has a success update and the agents' *lists* remain the same. At turn $t = 5$ a new speaker is drawn $s = 1$, it broadcasts $d_s = 0$. The listeners zero and three have a consistent element to this broadcast, therefore their K is not empty. Furthermore, listeners two has no consistent elements to put in K, and finds the rows two and three from its *variable domain* to be consistent to this broadcast. The overall outcome is a failure and thus we have a failure update. The listeners zero and two have empty K's so they do not change their *lists* and listener two adds two new elements in its lists. At turn $t = 6$ agent two speaks and broadcasts the element $d_s = 3$. Agent three does not have consistent elements to this broadcast and thus feedbacks a failure. Then we have a failure update and agent two adds two elements to its *list*. At turn $t = 7$ agent one speaks and broadcasts the element $d_s = 0$. All the listeners have consistent elements therefore their Ks are not empty. We get a success update. The agents two and three both delete an element from their *lists* which is not consistent to the speaker broadcast. At turn $t = 8$ agent one speaks again and broadcasts the same element $d_s = 0$. The system is unchanged. At turn $t = 9$ agent zero speaks and broadcasts the element $d_s = 2$. All listeners have consistent elements, therefore, there is a success update. Listener two deletes an element which was not consistent with the speaker broadcast. At turn $t = 10$ agent one speaks and broadcasts the element $d_s = 0$. All listeners have consistent elements to this broadcast, there is a success update, and the system is unchanged. At turn $t = 11$ agent zero speaks and broadcasts the element $d_s = 2$. All listeners have consistent elements to this broadcast, there is a success update, and the system is unchanged. At turn $t = 12$ agent three speaks and broadcasts the element $d_s = 1$. All listeners have consistent elements to this broadcast, there is a success update. Since the speaker had a different element in its list from the broadcasted element $d_s = 1$, he deletes this other element from its *list*. At this point all the elements in the agents *lists* are mutually consistent. Therefore, all the successive turns will have success updates and the system will not change any more. The system has found its global equilibrium state which is a solution of the puzzle we intended to solve.

3.1   Difference with Prior Self-stabilizing DCSPs

An agent in our algorithm is a finite state-machine. The agent (finite state-machine) evolution in time is represented by a *transition function* which depends on its state and its neighbors' states in the current turn. In particular the communication outcome is determined by the local state, where for local state s_1 we consider the agent state and its neighbors' state all together.

The communication outcome in the prior DCSPs can be forecast by the *transition function*. The *transition function* of an agent, which state is a_i, in the local state s_i is one and one only, and we can forecast exactly its next state a_{i+1} and the next local state s_{i+1}.

Let uniform protocols be distributed protocols, in which all the nodes are logically equivalent and identically programmed. It has been proved that in particular situations uniform self-stabilizing algorithms can not always solve the CSPs ([6]), in particular if we consider the ring ordering problems. In ring ordering problems we have N numbered nodes $\{n_1, n_2, \ldots, n_N\}$ ordered on a cycle graph. Each node has a variable, the variable assignment of the $i + 1$-th node n_{i+1} is the consecutive number of the variable assignment of the i-th node n_i in modulo N. The *variable domain* is $\{0, 1, \ldots, N - 1\}$ and every link has the constraint $\{n_i = j, n_{i+1} = (j + 1) \mod N | 0 \leq j \leq N\}$. Dijkstra [2] and Collin [6] propose dropping the uniform protocol condition to make the problem solvable.

Our protocol overcomes this by introducing random behavior. Moreover, the agent state is defined by an array that attributes a zero or a one to each element of the agent domain. The array element will be zero if the element is not in the *list*, and one if the element is in the list. This array determines a binary number a_i, which defines the state of the agent. If we know the states of all the agents, the transition of each agent from state a_i to a_j is uniquely determined once we know the agent that will be the speaker, the agents that will be the listeners, and the element that will be broadcasted by the speaker (Fig.2(a)).

Since the speaker will be chosen randomly, we can compute the probability for each agent being a speaker P_s. From this information, since we know the underling graph and that all his neighbors will be listeners, we can compute the probability for each agent to be a listener P_l. Knowing the speaker state, we can compute the probability for each element to be broadcast P_b. At this point we may be able to compute the *probabilistic transition function* $T(P_s, P_l, P_b)$, which will depend on the probabilities that we have just defined (Fig.2(b)).

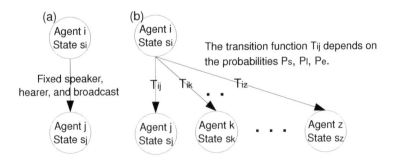

Fig. 2. (a) Shows that once we determine the speaker state, the broadcast, and the listener state we are able to determine the speaker and listeners' transitions. (b) Shows that since we have the speaker probability P_s, the broadcast probability P_b, and the listener probability P_l we can determine the *probabilistic transition function* $T(P_s, P_l, P_b)$.

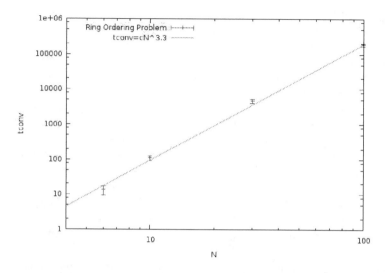

Fig. 3. The plot shows the ring ordering problem with N nodes, we see for the convergence turn t_{conv} the scaling proportion: $t_{conv} \propto N^{3.3}$

In this setting the agent state a_t at turn t can now be represented by a discrete distribution function and the *transition function* is now a Markovian Chain, the arguments of which are the transition probabilities p_j between the local states s_i and s_j. Thus we speak of a *probabilistic transition function*, which represents the probability of finding the system in a certain state s_j starting from s_i at time t. This behavior induces the algorithm to explore the state space randomly, until it finds the stable state that represents our expected solution. In the following plot we show the *convergence turn* t_{conv} scaling <u>with</u> the size N of the ring ordering problem. We average the *convergence turn* t_{conv} on ten runs of our algorithm for a set size N. Then we plot this point in a double logarithmic scale to evince the power law exponent of the function. We found that $t_{conv} \propto N^{3.3}$.

3.2   Analytical Description

In this section we are going probabilistically analyze how our algorithm solves graph problems for the following graphs structures: path graph and completely connected graph. These are simple limiting cases that help us to picture how our algorithm evolves in more general cases.

Path graph. The way our algorithm solves a path graph coloring instance can be described through analytical consideration; similar observations can then be extended to the cycle graph. The system dynamics are analogous to the naming game on a one dimensional network [8]. To each node of the path graph we attribute a natural number in increasing order, from the first node of the path, to which we attribute 1, to the last node, to which we attribute N. We can see that from a global point of view there are two final states: one state with odd number nodes of one color and even number nodes of the second color; the other state is inverted. At the beginning, when $t < N/3$, the system is dominated by new local consistent nodes' neighborhoods, which emerge sparsely and propagate to the connected nodes. The speaker has an empty *list* and it has to draw the assignment from the two element *variable domain*. In this way it selects one of the two final states to which it starts to belong. By communicating with its neighbors it makes them choose the same final state. We have in this way a small consistent domain of three agents that agree on the final state.

Since the speakers are chosen by the scheduler randomly, after some time, $t > N/3$, all the agents have been speakers or listeners at least once. Thus we find approximately $N/3$ domains dispersed in little clusters of generally three agents. Each of these domains belong to one final state or to the other.

At this point the domains start to compete. Between two domains we see an overlapping region appear. This region is constituted by agents that have more than one element in their lists. We can refer to them as undecided agents that belong to both domains, since the agents are on a path graph this region is linear. By probabilistic consideration we can see that this region tends to enclose less than two agents. For this reason we define the region that they form as a *border*, for a path graph of large size N the border width is negligible. So we approximate that only one agent is within this *border*. Under this hypothesis we can evaluate the evolution of the system as a diffusion problem, in which the borders move in a random walk on the path graph. When a domain grows over another domain, the second domain disappears. Thus the relation between the cluster growth and time is $\Delta x \propto (\frac{t}{\xi})^{1/2}$, where ξ is the time needed for the random walk to display a deviation of ± 1 step from its position. The probability that the border will move one step right or left on the path graph is $\propto 1/N$, proportional to the probability that an agent on the border or next to the border is extracted. Thus we can fix the factor $\xi \propto 1/N$. Since the lattice is long N we find the following relation for the average convergence turn $t_{conv} \propto N^3$. The average convergence turn t_{conv} is the average time at which the system finds global consistency. To calculate this we add the weighted convergence turns of all the algorithm runs, where the weights are the probabilities of the particular algorithm run.

Completely connected graph. Since all the variables in the graph coloring of a completely connected graph are bound by a inequality constraint, these variables must all be different. Thus N colors are necessary in this graph. To color the completely connected graph the agents start with a color domain of cardinality N.

At the beginning all the agents' lists are empty. The first speaker chooses a color and since all the agents are neighbors, it communicates with all of them. The listener selects the colors from the *variable domain* consistent to the color picked by the speaker. In the following turns the interactions are always successful. Two cases may be observed: the speaker has never spoken so it selects a color from its list and it shows the choice while the listeners cancel the same color from their lists; or the speaker has already spoken once, so there are no changes in the system because it already has only one color and all the other agents have already deleted this assignment. Since at each turn only one agent is a speaker, to let the system converge all the agents have to speak once.

Let N be the number of agents and $n(t-1)$ the number of agents that have spoken once before turn t. The probability that a new agent will speak at turn t considering that $n(t-1)$ agents have spoken already at turn $t-1$ is:

$$P(X_n^{(t)} = 1) = 1 - (\frac{n}{N}) \tag{1}$$

$$P(X_n^{(t)} = 0) = \frac{i}{N}. \tag{2}$$

Where $X_n^{(t)}$ is a random variable that is equal to one, when a new agent speaks at turn t, and equal to zero, when the agent that speaks at turn t has already spoken once. We calculate the probability that all agents have spoken once at a certain turn t, this is the probability $P(t_conv)$ that the system converges at turn t. We use this probability to compute the weighted average turn at which the system converges. This will be the weighted average convergence turn t_{conv}. We calculate this average by calculating the absorbing time of the corresponding absorbing Markow chain problem. If we consider as the beginning state, the state in which no agent has spoken $n(t=0) = 0$, we find the convergence time:

$$t_{conv} = N \sum_{j=1}^{N-1} \frac{1}{N-j} = N \sum_{k=1}^{N-1} \frac{1}{k} \sim N \log(N-1) \tag{3}$$

This, as we stated above corresponds to the time of convergence of our system, when it is trying to color a completely connected graph.

3.3   Algorithm Test

We have tested the above algorithm in the following classical CSP problems: graph coloring and n-queens puzzle. We plotted the graph of the convergence turn t_{conv} scaling with the number N of the CSP variables, each point was measured by ten runs of our algorithm. We considered four types of graphs for the graph coloring: path graphs, cycle graphs, completely connected graphs, and Mycielsky graphs.

N-Queens Puzzle. In the case of the N-queens puzzle with N variables, we measure the scaling proportion $t_{conv} \propto N^{4.2}$ for the convergence turn (Fig. 4).

Graph Coloring. In the study of graph coloring we presented four different graph structures:

- path graphs
- cycle graphs
- completely connected graph
- Mycielsky graphs.

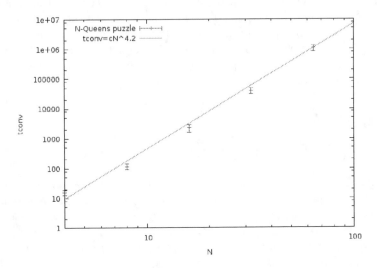

Fig. 4. The plot shows the N-queens puzzle with N variables, we see for the convergence turn t_{conv} the scaling proportion: $t_{conv} \propto N^{4.2}$. The points on this graph are averaged on ten algorithm runs.

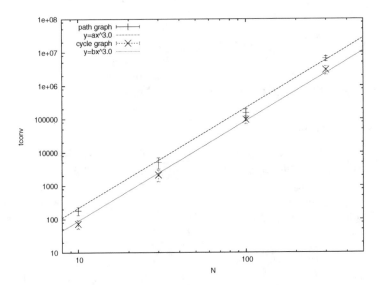

Fig. 5. The plot shows the graph coloring in the case of path graphs and special 2-colorable cycle graphs with 2 colors. The convergence turn t_{conv} of the path graphs and cycle graphs exhibit a power law behavior $t_{conv} \propto N^{3.0}$. The cycle graph exhibits a faster convergence. The points on this graph are averaged on ten algorithm runs.

In the study of the path graph and the cycle graph we have restricted ourselves to the $2 - chromatic$ cases: all the path graphs and only the even number node

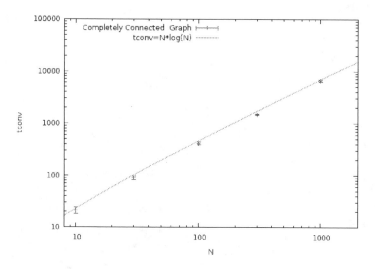

Fig. 6. The plot shows the graph coloring in the case of a completely connected graph with N colors: in this case we find that the convergence turn is $t_{conv} \sim N \log(N)$. The points on this graph are averaged on ten algorithm runs.

cycle graphs. Thus we imposed the agent variable domain to two colors. In this context the convergence turn t_{conv} of the path graph and the cycle graph exhibit a power law behavior $t_{conv} \propto N^{3.0}$. The cycle graph exhibits a faster convergence (Fig. 5). We see from these measurements that the power law of the convergent time scaled with the number of nodes N is compatible with our analytical considerations.

The graph coloring in the case of a completely connected graph always needs at least N colors: in this case we find that the convergence turn is $t_{conv} \propto N \log(N)$ (Fig. 6).

The Mycielski graph [15] of an undirected graph G is generated by the Mycielski transformation on the graph G and is denoted as $\mu(G)$ (see Fig.7). Let the N number of nodes in the graph G be referred to as $v_1, v_2, \ldots, v_N$. The Mycielski graph is obtained by adding to graph G $N+1$ nodes: N of them will be named $u_1, u_2, \ldots, u_N$ and the last one w. We will connect with an edge all the nodes $u_1, u_2, \ldots, u_N$ to w. For each existing edge of the graph G between two nodes v_i and v_j we include an edge in the Mycielski graph between v_i and u_j and between u_i and v_j.

The Mycielski graph of graph G of N nodes and E edges has $2N+1$ nodes and $2E+N$ edges.

Iterated Mycielski transform applications starting from the null graph, generates the graphs $M_i = \mu(M_{i-1})$. The first graphs of the sequence are M_0 the null graph, M_1 the one node graph, M_2 the two connected nodes graph, M_3 the five nodes cycle graph, and M_4 the Grötzsch graph with 11 vertices and 20 edges

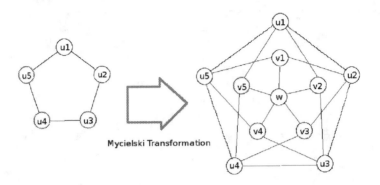

Fig. 7. Mycielski transformation of a five node cycle graph

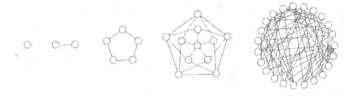

Fig. 8. Mycielski graph sequence M_1, M_2, M_3, M_4, and M_5[16]

Table 1. Convergence turn t_{conv} of the Mycielski graph coloring. M_i is the Mycielski graph identification, N is the number of nodes, E is the number of edges, k the optimal coloring, and t_{conv} the convergence turn.

M_i	N	E	k optimal coloring	t_{conv}
M_4	11	20	4	32 ± 2
M_5	23	71	5	170 ± 20
M_6	47	236	6	3300 ± 600
M_7	95	755	7	$(1.1 \pm 0.2) \cdot 10^6$

(see Fig. 8). The number of colors k needed to color a graph M_i of the Mycielski sequence is, $k = 1$ ([15]).

These graphs are particularly difficult to solve because they do not possess triangular cliques, moreover, they have cliques of higher order and the coloring number increases each Mycielski transformation ([14]). We ran our algorithm to solve the graph coloring problem with the known optimal coloring. Table 1 shows for each graph of the Mycielski sequence M_i, the number of nodes N, the number of edges E, the minimal number of colors needed k and the convergence turn t_{conv} of our algorithm.

4   Conclusions and Future Work

Our aim is to develop a probabilistic algorithm able to find the solution of a CSP instance. In the study of this method we are trying to fully exploit the power of distributed calculation. To do this we generalize the naming game algorithm, by letting the CSP solution emerge, rather than being the conclusion of a sequence of statements. As we saw in Subsection 3.2 our algorithm is based on the random exploration of the system state space. Our algorithm travels through the possible states until it finds the absorbing state, where it stabilizes. These ergodic features guarantee that the system to has a probability equal to one to converge [13] for long times $t \to +\infty$. Unfortunately this time depending on the particular CSP instance can be too long for practical use.

This is achieved through the union of new topics addressed in statistical physics (the naming game), and the abstract framework posed by constraint solving.

In future work we will test the algorithm on a uniform random binary CSP to fully validate this method. We also expect to generalize the communication model to let more then one agent speak at the same turn. Once we have done this we can let the agents speak spontaneously without a central scheduler.

References

1. Gosti, G.: Resolving CSP with Naming Games. In: Garcia de la Banda, M., Pontelli, E. (eds.) ICLP 2008. LNCS, vol. 5366, pp. 807–808. Springer, Heidelberg (2008)
2. Dijkstra, E.W.: Self-Stabilizing Systems in Spite of Distributed Control. Communications of the ACM 17(11), 643–644 (1974)
3. Ramanathan, R., Lloyes, E.L.: Scheduling Algorithms for Multi-Hop Radio Networks. In: Proceedings of the SIGCOMM 1992, Communication Architectures and Protocols, pp. 211–222. ACM Press, New York (1992)
4. Lessr, V.R.: An Overview of DAI: Viewing Distributed AI as Distributed Search. Japanese Society for Artificial Intelligence 5(4) (1990)
5. Yokoo, M., Durfee, E.H., Ishida, T., Kuwabara, K.: Distributed Constraint Satisfaction for Formalizing Distributed Problem Solving. In: 12th International Conference on Distributed Computing Systems (ICDCS 1992), pp. 614–621 (1992)
6. Collin, Z., Dechter, R., Katz, S.: On the Feasibility of Distributed Constraint Satisfaction. In: Proceedings of the Twelfth International Joint Conference of Artificial Intelligence (IJCAI 1991) (1991)
7. Baronchelli, A., Felici, M., Caglioti, E., Loreto, V., Steels, L.: Sharp Transition Toward Shared Vocabularies in Multi-Agent Systems. Journal of Statistical Mechanics, P06014 (2006)
8. Baronchelli, A., Dall'Asta, L., Barrat, A., Loreto, V.: Topology Induced Coarsening in Language Games Language. Phys. Rev. E 73, 015102(R) (2006)
9. Steels, L.: Self-Organizing Vocabularies. In: Langton, C., Shimohara, K. (eds.) Artificial Life V: Proceeding of the Fifth International Workshop on the Synthesis and Simulation of Living Systems, pp. 179–184 (1996)
10. Nowak, M.A., Plotkin, J.B., Krakauer, J.D.: The Evolutionary Language Game. Journal of Theoretical Biology 200, 147 (1999)

11. Nowak, M.A., Komarova, N.L., Niyogi, P.: Computational and Evolutionary Aspects of Language. Nature 417, 611–617 (2002)
12. Lenaerts, T., Jansen, B., Tuyls, K., de Vylder, B.: The Evolutionary Language Game: An orthogonal approach. Journal of Theoretical Biology 235(4), 566–582 (2005)
13. Grinstead, C.M., Snel, J.L.: Introduction to Probability. American Mathematical Society, Providence (2003)
14. Trick, M.: Network Resources for Coloring a Graph,
    http://mat.gsia.cmu.edu/COLOR/color.html
15. Mycielski, J.: Sur le coloriage des graphes. Colloq. Math. 3, 161–162 (1955)
16. Weisstein, E.W.: Mycielski Graph. MathWorld–A Wolfram Web Resource,
    http://mathworld.wolfram.com/MycielskiGraph.html
17. Leighton, F.T.: A Graph Coloring Algorithm for Large Scheduling Problems. Journal of Research of the National Bureau of Standards 84, 489–505 (1979)

An Efficient Decision Procedure for Functional Decomposable Theories Based on Dual Constraints

Khalil Djelloul

Laboratoire d'Informatique Fondamentale d'Orléans,
Bat. 3IA, rue Léonard de Vinci. 45067 Orléans, France

Abstract. Over the last decade, first-order constraints have been efficiently used in the artificial intelligence world to model many kinds of complex problems such as: scheduling, resource allocation, computer graphics and bio-informatics. Recently, a new property called *decomposability* has been introduced and many first-order theories have been proved to be decomposable such as finite or infinite trees, rational and real numbers, linear dense order,...etc. A decision procedure in the form of five rewriting rules has also been developed. It decides if a first-order formula without free variables (proposition) is true or not in any decomposable theory. Unfortunately, this later needs to normalize the initial proposition before starting the solving process. This transformation generates many nested negations and quantifications which greatly slow down the performances of this decision procedure. We present in this paper an efficient decision procedure for functional decomposable theories, i.e. theories whose set of relation is reduced to $\{=, \neq\}$. This new decision procedure does not need to normalize the formulas and transforms any first-order proposition with any logical symbols into a boolean combination of basic formulas which are either equivalent to true or to false. We show the efficiency of our algorithm (in time and space) and compare its performances with those of the classical decision procedure for decomposable theories. Our algorithm is able to solve first-order propositions involving many nested alternated quantifiers of the form $\exists \bar{x} \forall \bar{y}$ over different functional decomposable theories.

1 Introduction

First-order constraints are first-order formulas built on a set of function and relation symbols using the following logical symbols: $=, \neq, true, false, \neg, \wedge, \vee, \rightarrow, \leftrightarrow,$ $\forall, \exists, (,)$. Over the last decade, first-order constraints have been efficiently used in the artificial intelligence world to model many kinds of complex problems such as: scheduling, resource allocation, configuration, temporal and spatial reasoning, computer graphics, bio-informatics [1,10]. However, in most of the cases, the quantifiers are not used due to the inherent huge complexity in time and space when solving first-order constraints with imbricated quantifiers, such as:

$$\exists x \forall y \begin{bmatrix} x = f(y,x) \wedge f(x, f(w,y)) = f(f(y,x), w) \wedge \\ \neg(\forall v \exists z \, (x = f(v,x) \rightarrow w = f(z,w))) \end{bmatrix},$$

A. Oddi, F. Fages, and F. Rossi (Eds.): CSCLP 2008, LNAI 5655, pp. 33–50, 2009.
© Springer-Verlag Berlin Heidelberg 2009

and if we use Maher's theory of finite or infinite trees [6,11] then solving such a constraint cannot be done with an algorithm of better complexity in time and space than a huge tower of powers of two, i.e. $2^{2^{2^{\cdots}}}$ whose depth is proportional to the number of imbricated quantifiers [3,17]. Due to this high complexity, only few general first-order constraint solvers have been developed in the past and no one of them could solve complex first-order constraints with many imbricated quantifiers.

Recently, we showed that a lot of first-order theories such as: finite or infinite trees, real numbers, rational numbers, linear dense order without endpoints,...etc share a new property that we called *decomposability* [7]. We have then presented a decision procedure in the form of five rewriting rules which for any decomposable theory T can decide the satisfiability or unsatisfiability of any first-order proposition, i.e. any first-order constraint whose all variables are quantified, such as:

$$\exists u_2 \forall u_1 \exists u_3 \neg \begin{bmatrix} \exists v_1\, v_1 = f(u_1, u_2) \wedge u_2 = g(u_1) \wedge \\ \neg(\exists w_1\, v_1 = g(w_1)) \wedge \\ \neg(\exists w_2\, u_2 = g(w_2) \wedge w_2 = g(u_3)) \end{bmatrix}.$$

In order to decide the truth value of any proposition φ, our decision procedure uses a pre-processing step which transforms φ into a particular form of formulas called *normalized formulas*, i.e. formulas of the form

$$\neg(\exists x_1 ... \exists x_n\, (\alpha \wedge \bigwedge_{i \in I} \varphi_i)), \tag{1}$$

with α a conjunction of atomic formulas and the $\varphi_i's$ sub-normalized formulas of the same form an (1). Let us choose for instance the theory Tr of finite or infinite trees [7,11]. Let f and g be two function symbols. The following formula is normalized:

$$\neg \left[\exists y\, y = f(y) \wedge \begin{bmatrix} \neg(\exists x\, y = f(x) \wedge x = g(y) \wedge \neg(\exists z\, z = g(y))) \wedge \\ \neg(\exists v\, v = f(y)) \end{bmatrix} \right].$$

Once the normalized formula is obtained, our decision procedure can be used. It transforms any normalized proposition into true or false. The problem is that, in most of the cases, the normalized formula ϕ obtained from the initial proposition φ is greatly bigger (more nested quantifiers and negations) than φ. In fact, if for example we use the theory Q of rational numbers and if φ is the formula

$$\forall x \exists y\, y \neq x,$$

then ϕ is the following normalized formula

$$\neg(\exists x\, true \wedge \neg(\exists y\, true \wedge \neg(\exists \epsilon\, y = x))),$$

where $\exists \epsilon$ is the empty existential quantification. In this example, we move from a very simple formula into a complex one with three nested imbricated quantifiers and negations. By using our decision procedure on ϕ we get an execution time which is larger than the one obtained using a direct simplification of φ into *true* since for all variable x we can find a variable y which is different from x in Q. This big difference between the two execution times is due to the fact that each time we have two imbricated negations of quantifiers, the decision procedure of [7] uses a very costly distribution rule which eliminates one level of nested negations and quantifications but exponentially increases the size of the formula, and so on until reaching the formula *true* or the formula *false*. In other words, the more nested quantifiers and negations we have in the normalized formula ϕ, the higher its execution time will be. It is then much more interesting for us if we can compute the truth value directly from the initial constraint φ without transforming it into a normalized formula ϕ since normalizing a formula implies the generation of many nested negations and quantifiers. Unfortunately, there exists so far no algorithm for decomposable theories which does not use the normalized formulas.

Contributions. In this paper, we build a new efficient decision procedure for functional decomposable theories, i.e. decomposable theories whose signature does not contain relations other than $=$ and $\neq$[1]. Our new algorithm does not use any particular form of formulas and can be applied to any first-order formula φ with any logical and functional symbols. It does not need to transform the initial constraint φ into a normalized formula and uses a new approach as well as new properties which are completely different from those used by our old decision procedure for decomposable theories [7].

The main idea behind our algorithm consists in using new properties of functional decomposable theories which enables us to not transform the initial formula φ into a normalized one and to build directly from φ a boolean combination of particular formulas which can be immediately reduced to *true* or to *false*. To this end, we introduce the so called *dual* and *basic* formulas and build a set of rewriting rules which handles dual formulas and transforms any proposition φ into a boolean combination of basic formulas which can be immediately reduced to *true* or to *false*.

This paper is organized in four sections followed by a conclusion. This introduction is the first section. Section 2 is dedicated to a brief recall on first-order logic and decomposable theories. We present in Section 3 our decision procedure given in the form of 18 rewriting rules. We end this paper by a series of benchmarks realized by a C++ implementation of our algorithm over two functional decomposable theories. We show that our decision procedure is much more efficient in time and space than the classical decision procedure for decomposable theories. The dual formulas, the working formulas, the algorithm and the benchmarks are our new contributions in this paper.

2   Preliminaries

2.1   First-Order Formulas

Let V be an infinite set of variables. Let S be a set of symbols, called a signature and partitioned into two disjoint sub-sets: the set F of function symbols and

---

[1] Of course, these theories can have any set of function symbols.

the set R of relation symbols. To each function symbol and relation is linked a non-negative integer n called its *arity*. An n-ary symbol is a symbol of arity n. A first-order constraint or formula is an expression of one of the eleven following forms:

$$s = t, \ r(t_1, \ldots, t_n), \ true, \ false,$$
$$\neg\varphi, \ (\varphi \wedge \psi), \ (\varphi \vee \psi), \ (\varphi \rightarrow \psi), \ (\varphi \leftrightarrow \psi), \tag{2}$$
$$(\forall x \, \varphi), \ (\exists x \, \varphi),$$

with $x \in V$, r an n-ary relation symbol taken from R, φ and ψ shorter formulas, s, t and the t_is terms, that are expressions of the one of the following two forms $x, f(t_1, \ldots, t_n)$, with x taken from V, f an n-ary function symbol taken from F and the t_i's shorter terms. The formulas of the first line of (2) are known as *atomic*, and *flat* if they are of one of the following forms:

$$true, \ false, \ x_0 = x_1, x_0 = f(x_1, \ldots, x_n), \ r(x_1, \ldots, x_n),$$

with the x_i's (possibly non-distinct) variables taken from V, $f \in F$ and $r \in R$. We denote by AT the set of the conjunctions of flat atomic formulas.

An occurrence of a variable x in a formula ϕ is *bound* if it occurs in a sub-formula of the form $(\forall x \, \varphi)$ or $(\exists x \, \varphi)$. It is *free* in the contrary case. The *free variables of a formula* are those which have at least one free occurrence in this formula. A *proposition* is a formula without free variables.

A *model* is a pair $M = (D, F)$, where D is a non-empty set of individuals of M and F a set of functions and relations in D. We call *instantiation* of a formula φ by individuals of M, the formula obtained from φ by replacing each free occurrence of a free variable x in φ by the same individual i of D and by considering each element of D as 0-ary function symbol.

A *theory* T is a (possibly infinite) set of propositions. We say that the model M is a *model of* T, if for each element φ of T, $M \models \varphi$. If φ is a formula, we write $T \models \varphi$ if for each model M of T, $M \models \varphi$. A theory T is *complete* if for every proposition φ, one and only one of the following properties holds: $T \models \varphi$, $T \models \neg\varphi$.

2.2   Vectorial Quantifiers

Let M be a model and T a theory. Let $\bar{x} = x_1 \ldots x_n$ and $\bar{y} = y_1 \ldots y_n$ be two words on V of the same length. Let φ, and $\varphi(\bar{x})$ be formulas. We write

$$\begin{array}{ll} \exists \bar{x} \, \varphi & \text{for } \exists x_1 \ldots \exists x_n \, \varphi, \\ \forall \bar{x} \, \varphi & \text{for } \forall x_1 \ldots \forall x_n \, \varphi, \\ \exists? \bar{x} \, \varphi(\bar{x}) & \text{for } \forall \bar{x} \forall \bar{y} \, \varphi(\bar{x}) \wedge \varphi(\bar{y}) \rightarrow \bigwedge_{i \in \{1, \ldots, n\}} x_i = y_i, \\ \exists! \bar{x} \, \varphi & \text{for } (\exists \bar{x} \, \varphi) \wedge (\exists? \bar{x} \, \varphi). \end{array}$$

The word $\bar{x}$, which can be the empty word ε, is called *vector of variables*. Note that semantically the new quantifiers $\exists?$ and $\exists!$ simply means "at most one" and "one and only one".

Let us now introduce a convenient notation concerning the priority of the quantifiers: $\exists$, $\exists!$, $\exists?$ and $\forall$.

Notation 2.2.1. *Let Q be a quantifier taken from $\{\forall, \exists, \exists!, \exists?\}$. Let $\bar{x}$ be vector of variables taken from V. We write:*

$$Q\bar{x}\,\varphi \wedge \phi \quad for \quad Q\bar{x}\,(\varphi \wedge \phi).$$

Example 1. Let $I = \{1, ..., n\}$ be a finite set with $n \geq 0$. Let φ and ϕ_i with $i \in I$ be formulas. Let $\bar{x}$ and $\bar{y}_i$ with $i \in I$ be vectors of variables. We write:

$$
\begin{array}{ll}
\exists \bar{x}\,\varphi \wedge \neg\phi_1 & \text{for } \exists \bar{x}\,(\varphi \wedge \neg\phi_1), \\
\forall \bar{x}\,\varphi \wedge \phi_1 & \text{for } \forall \bar{x}\,(\varphi \wedge \phi_1), \\
\exists!\bar{x}\,\varphi \wedge \bigwedge_{i \in I}(\exists \bar{y}_i \phi_i) & \text{for } \exists!\bar{x}\,(\varphi \wedge (\exists \bar{y}_1 \phi_1) \wedge ... \wedge (\exists \bar{y}_n \phi_n) \wedge true), \\
\exists?\bar{x}\,\varphi \wedge \bigwedge_{i \in I} \neg(\exists \bar{y}_i \phi_i) & \text{for } \exists?\bar{x}\,(\varphi \wedge (\neg(\exists \bar{y}_1 \phi_1)) \wedge ... \wedge (\neg(\exists \bar{y}_n \phi_n)) \wedge true).
\end{array}
$$

Let us end this sub-section by two properties which will help us to prove the correctness of our decision procedure.

Property 2.2.2. *If $T \models \exists?\bar{x}\,\varphi$ then*

$$T \models (\exists \bar{x}\,\varphi \wedge \bigwedge_{i \in I} \neg\phi_i) \leftrightarrow ((\exists \bar{x}\varphi) \wedge \bigwedge_{i \in I} \neg(\exists \bar{x}\,\varphi \wedge \phi_i)).$$

Property 2.2.3. *If $T \models \exists!\bar{x}\,\varphi$ then*

$$T \models (\exists \bar{x}\,\varphi \wedge \bigwedge_{i \in I} \neg\phi_i) \leftrightarrow \bigwedge_{i \in I} \neg(\exists \bar{x}\,\varphi \wedge \phi_i).$$

2.3 The Infinite Quantifier $\exists_{\infty}^{\Psi(u)}$

Before introducing the definition of decomposable theories, we recall the definition of the infinite quantifier $\exists_{\infty}^{\Psi(u)}$ that we have given in [7]:

Definition 2.3.1. *[7] Let M be a model, $\varphi(x)$ a formula and $\Psi(u)$ a set of formulas having at most one free variable u. We write*

$$M \models \exists_{\infty}^{\Psi(u)} x\,\varphi(x), \tag{3}$$

if for every instantiation $\exists x\,\varphi'(x)$ of $\exists x\,\varphi(x)$ by individuals of M and for every finite subset $\{\psi_1(u), .., \psi_n(u)\}$ of elements of $\Psi(u)$, the set of the individuals i of M such that $M \models \varphi'(i) \wedge \bigwedge_{j \in \{1,...,n\}} \neg\psi_j(i)$ is infinite.

This infinite quantifier holds only for models whose set of individuals is infinite. Note that if $\Psi(u) = \{false\}$ then (3) simply means that M contains an infinite set of individuals i such that $\varphi(i)$. Informally, the notation (3) states that there exists a full elimination of quantifiers in formulas of the form $\exists x\,\varphi(x) \wedge \bigwedge_{j \in \{1,...,n\}} \neg\psi_j(x)$ due to an infinite set of valuations of x in M which satisfy this formula.

Property 2.3.2. *[7] Let J be a finite (possibly empty) set. Let $\varphi(x)$ and $\varphi_j(x)$ with $j \in J$ be M-formulas. If $T \models \exists_{\infty}^{\Psi(u)} x\,\varphi(x)$ and if for each $\varphi_j(x)$, at least one of the following properties holds:*

- $T \models \exists?x\, \varphi_j(x)$,
- there exists $\psi_j(u) \in \Psi(u)$ such that $T \models \forall x\, \varphi_j(x) \to \psi_j(x)$,

then

$$T \models \exists x\, \varphi(x) \wedge \bigwedge_{j \in J} \neg\varphi_j(x)$$

Property 2.3.3. *[7] If $T \models \exists_\infty^{\Psi(u)} x\, \varphi(x)$ then $T \models \exists_\infty^{\Psi(u)} x\, true$.*

2.4 Decomposable Theories

We now recall the definition of *decomposable theories* [7]. Informally, this definition simply states that in every decomposable theory T each formula of the form $\exists \bar{x}\, \alpha$, with $\alpha \in AT$, is equivalent in T to a decomposed formula of the form $\exists \bar{x}'\, \alpha' \wedge (\exists \bar{x}''\, \alpha'' \wedge (\exists \bar{x}'''\, \alpha'''))$ where the formulas $\exists \bar{x}'\, \alpha'$, $\exists \bar{x}''\, \alpha''$, and $\exists \bar{x}'''\, \alpha'''$ have elegant properties which can be expressed using the following quantifiers: $\exists?$, $\exists!$ and $\exists_\infty^{\Psi(u)}$.

In all what follows, we will use the abbreviation wnfv for *"without new free variables"*. A formula φ is equivalent to a wnfv formula ψ in T means that $T \models \varphi \leftrightarrow \psi$ and ψ does not contain other free variables than those of φ.

Definition 2.4.1. *A theory T is called* decomposable *if there exists a set $\Psi(u)$ of formulas, having at most one free variable u, and three sets A', A'' and A''' of formulas of the form $\exists \bar{x}\, \alpha$ with $\alpha \in AT$ such that:*

1. *Every formula of the form $\exists \bar{x}\, \alpha \wedge \psi$, with $\alpha \in AT$ and ψ any formula, is equivalent in T to a wnfv decomposed formula of the form*

$$\exists \bar{x}'\, \alpha' \wedge (\exists \bar{x}''\, \alpha'' \wedge (\exists \bar{x}'''\, \alpha''' \wedge \psi)),$$

   with $\exists \bar{x}'\, \alpha' \in A'$, $\exists \bar{x}''\, \alpha'' \in A''$ and $\exists \bar{x}'''\, \alpha''' \in A'''$.
2. *If $\exists \bar{x}'\alpha' \in A'$ then $T \models \exists?\bar{x}'\, \alpha'$ and for each free variable y in $\exists \bar{x}'\alpha'$, at least one of the following properties holds:*
   - *$T \models \exists?y\bar{x}'\, \alpha'$,*
   - *there exists $\psi(u) \in \Psi(u)$ such that $T \models \forall y\, (\exists \bar{x}'\, \alpha') \to \psi(y)$.*
3. *If $\exists \bar{x}''\alpha'' \in A''$ then for each x_i'' of $\bar{x}''$ we have $T \models \exists_\infty^{\Psi(u)} x_i''\, \alpha''$.*
4. *If $\exists \bar{x}'''\alpha''' \in A'''$ then $T \models \exists!\bar{x}'''\, \alpha'''$.*
5. *If the formula $\exists \bar{x}'\alpha'$ belongs to A' and has no free variables then this formula is either the formula $\exists \varepsilon true$ or $\exists \varepsilon false$.*

In [7] many first-order theories have been proved to be decomposable such as: theory of finite or infinite trees [6,11], Clark equational theories [2], rational and real numbers with addition and subtraction [9] and many combinations based on these theories [8]. From the proof of the decomposability of these theories we can deduce their completeness using a decision procedure which for every proposition produces either true or false [7]. This later uses a pre-processing step which transforms the initial proposition φ into a particular form called *normalized formula*. We will see in the next subsection the inconvenience of such a transformation.

2.5   Normalized Formulas

Definition 2.5.1. *A normalized formula φ of depth $d \geq 1$ is a formula of the form*

$$\neg(\exists \bar{x}\, \alpha \wedge \bigwedge_{i \in I} \varphi_i),$$

with I a finite (possibly empty) set, $\alpha \in AT$ and the φ_i's normalized formulas of depth d_i with $d = 1 + \max\{0, d_1, ..., d_n\}$.

Example 2.5.2. *Let φ be the following formula*

$$\forall x \exists y\, y \neq x \wedge x = f(x), \tag{4}$$

where f is a 1-ary function symbol. The preceding formula is equivalent in any decomposable theory T to the following normalized formula ϕ of depth 3:

$$\neg(\exists x\, true \wedge \neg(\exists y\, true \wedge \neg(\exists \varepsilon\, y = x \vee x \neq f(x)))). \tag{5}$$

In order to solve ϕ, the decision procedure of [7] uses among other things a rule of the form

$$\neg\left[\begin{array}{l} \exists \bar{x}\, \alpha \wedge \varphi \wedge \\ \neg\left[\begin{array}{l} \exists \bar{y}'\, \beta' \wedge \\ \bigwedge_{i \in I} \neg(\exists \bar{z}'_i\, \delta'_i) \end{array}\right] \end{array}\right] \implies \left[\begin{array}{l} \neg(\exists \bar{x}\, \alpha \wedge \varphi \wedge \neg(\exists \bar{y}'\, \beta')) \wedge \\ \bigwedge_{i \in I} \neg(\exists \bar{x}\bar{y}'\bar{z}'_i\, \alpha \wedge \beta' \wedge \delta'_i \wedge \varphi) \end{array}\right]$$

this later transforms a normalized formula of depth 3 into a conjunction of normalized formulas of depth 2. However, each time this rule decreases one depth of the normalized formula, it builds a huge conjunction of new normalized formulas. In fact, we showed in [7] that this rule is the only one responsible for the exponential complexity in time and space of our decision procedure. On the other hand, the transformation of the formula φ (the formula (4)) into the normalized formula ϕ (the formula (5)) implies the creation of three nested negations and thus we must apply two times our costly rule in order to decrease the depth of the formula ϕ and get true or false. All these steps can be avoided by a direct simplification of the initial constraint φ into true or false using new properties of decomposable theories. This will be the scope of the next section of this paper.

3   A Decision Procedure for Functional Decomposable Theories Based on Dual Formulas

We will see in this section how to build a new decision procedure which does not need to transform the initial formula into a normalized formula. This algorithm can be used for any functional decomposable theory, i.e. decomposable theory whose set of relation is reduced to $\{=, \neq\}$.

3.1 Dual Formulas

Let T be a functional decomposable theory together with the following signature $F \cup \{=, \neq\}$ where F is a (possibly infinite) set of function symbols. The sets $\Psi(u)$, A, A', A'' and A''' are now known and fixed for all the following sections.

Definition 3.1.1. *A* working formula *is a first-order formula which does not contain any occurrence of the logical symbol* $\neg$.

It is clear that for any functional decomposable theory T we can transform any first-order formula φ into a working formula. For that it is enough to distribute the negation into the sub-formulas according to the classical rules of first-order logic.

Example 3.1.2. *Let* φ *be the following first-order formula*

$$\exists u_2 \forall u_1 \exists u_3 \neg \begin{bmatrix} \exists v_1 \, v_1 = f(u_1, u_2) \wedge u_2 = g(u_1) \wedge \\ \neg(\exists w_1 \, v_1 = g(w_1)) \wedge \\ \neg(\exists w_2 \, u_2 = g(w_2) \wedge w_2 = g(u_3)) \end{bmatrix}. \tag{6}$$

The preceding formula is equivalent to the following working formula

$$\exists u_2 \forall u_1 \exists u_3 \begin{bmatrix} \forall v_1 \, v_1 \neq f(u_1, u_2) \vee u_2 \neq g(u_1) \vee \\ (\exists w_1 \, v_1 = g(w_1)) \vee \\ (\exists w_2 \, u_2 = g(w_2) \wedge w_2 = g(u_3)) \end{bmatrix}. \tag{7}$$

Definition 3.1.3. *The* dual $\overline{\varphi}$ *of a working formula* φ, *is the formula obtained by replacing each occurrence of* $=$, $\neq$, $\wedge$, $\vee$, $\exists$, $\forall$ *by* $\neq$, $=$, $\vee$, $\wedge$, $\forall$, $\exists$.

Example 3.1.4. *The dual of the working formula (7) is the following formula*

$$\forall u_2 \exists u_1 \forall u_3 \begin{bmatrix} \exists v_1 \, v_1 = f(u_1, u_2) \wedge u_2 = g(u_1) \wedge \\ (\forall w_1 \, v_1 \neq g(w_1)) \wedge \\ (\forall w_2 \, u_2 \neq g(w_2) \vee w_2 \neq g(u_3)) \end{bmatrix}.$$

Property 3.1.5. *We show that* $T \models \varphi \leftrightarrow \overline{\overline{\varphi}}$ *and* $T \models \varphi \leftrightarrow \neg\overline{\varphi}$.

3.2 Basic Formulas

Definition 3.2.1. *A basic formula is a formula of the one of the following two forms*

$$(\exists \bar{x} \, \alpha), (\forall \bar{x} \, \overline{\alpha}),$$

with $\alpha \in AT$ *and* $\bar{x}$ *a (possibly empty) vector of variables.*

Using Definition 2.4.1, we show the following property:

Property 3.2.2. *If* φ *is a basic formula without free variables then it is of one of the following forms*

$$(\exists \varepsilon \, true), (\exists \varepsilon \, false), (\forall \varepsilon \, true), (\forall \varepsilon \, false).$$

Proof. Let $\exists \bar{x}\, \alpha$ be a basic formula with $\alpha \in AT$. According to Definition 2.4.1 this formula is equivalent in T to a wnfv formula of the form

$$\exists \bar{x}'\, \alpha' \wedge (\exists \bar{x}''\, \alpha'' \wedge (\exists \bar{x}'''\, \alpha''')),$$

with $\exists \bar{x}'\, \alpha' \in A'$, $\exists \bar{x}''\, \alpha'' \in A''$ and $\exists \bar{x}'''\, \alpha''' \in A'''$. Since $\exists \bar{x}'''\, \alpha''' \in A'''$ then according to Definition 2.4.1 we have $T \models \exists !\bar{x}'''\alpha'''$, thus $T \models \exists \bar{x}'''\alpha'''$. The preceding formula is then equivalent in T to

$$\exists \bar{x}'\, \alpha' \wedge (\exists \bar{x}''\, \alpha''),$$

which is equivalent in T to

$$\exists \bar{x}'\, \alpha' \wedge (\exists x_1''...x_{n-1}''\, (\exists x_n''\, \alpha'')).$$

Since $\exists \bar{x}''\, \alpha'' \in A''$ then according to Definition 2.4.1 we have $T \models \exists_\infty^{\Psi(u)} x_n''\, \alpha''$ and thus according to Property 2.3.2 (with $J = \emptyset$) $T \models \exists x_n''\, \alpha''$. The preceding formula is equivalent in T to

$$\exists \bar{x}'\, \alpha' \wedge (\exists x_1''...x_{n-1}''\, true),$$

which is finally equivalent in T to

$$\exists \bar{x}'\, \alpha'.$$

According to the fifth point of Definition 2.4.1 the preceding formula is either the formula $\exists \varepsilon\, true$ or $\exists \varepsilon\, false$. By following the same steps and using Property 3.1.5, we show the rest of this property for the basic formulas of the form $\forall \bar{x}\, \overline{\alpha}$.

3.3 The Decision Procedure

Let φ be a proposition. Computing the truth value of φ in T proceeds as follows:
(1) Transform φ into a working formula ϕ.
(2) Apply the rewriting rules bellow on a sub-working formula of ϕ by considering that the connectors $\wedge$ and $\vee$ are associative and commutative.
(3) Repeat the second step until no rule can be applied. We get at the end either the formula $true$ or the formula $false$.

Distribution

$$
\begin{aligned}
&(1) \;\; \exists \bar{x}\, \varphi_1 \wedge (\varphi_2 \vee \varphi_3) \Rightarrow \exists \bar{x}\, (\varphi_1 \wedge \varphi_2) \vee (\varphi_1 \wedge \varphi_3) \\
&(2) \;\; \forall \bar{x}\, \varphi_1 \vee (\varphi_2 \wedge \varphi_3) \Rightarrow \forall \bar{x}\, (\varphi_1 \vee \varphi_2) \wedge (\varphi_1 \vee \varphi_3) \\
&(3) \;\; \exists \bar{x}\, \varphi_1 \vee \varphi_2 \quad\quad\;\; \Rightarrow \exists \bar{x}\, \varphi_1 \vee \exists \bar{x}\, \varphi_2 \\
&(4) \;\; \forall \bar{x}\, \varphi_1 \wedge \varphi_2 \quad\quad\;\; \Rightarrow \forall \bar{x}\, \varphi_1 \wedge \forall \bar{x}\, \varphi_2
\end{aligned}
$$

Lifting quantifications

$$
\begin{aligned}
&(5) \;\; \exists \bar{x}'\, \alpha' \wedge (\exists \bar{y}'\beta' \wedge \bigwedge_{i \in I}(\forall \bar{z}'\, \overline{\lambda}_i')) \wedge \varphi_1 \Rightarrow \exists \bar{x}'\bar{y}'\, \alpha' \wedge \beta' \wedge \bigwedge_{i \in I}(\forall \bar{z}'\, \overline{\lambda}_i') \wedge \varphi_1 \\
&(6) \;\; \forall \bar{x}'\, \overline{\alpha}' \vee (\forall \bar{y}'\overline{\beta}' \vee \bigvee_{i \in I}(\exists \bar{z}'\, \lambda_i')) \vee \varphi_1 \Rightarrow \forall \bar{x}'\bar{y}'\, \overline{\alpha}' \vee \overline{\beta}' \vee \bigvee_{i \in I}(\exists \bar{z}'\, \lambda_i') \vee \varphi_1
\end{aligned}
$$

Local solving

$$(7)\ \exists \overline{x}\, \alpha \wedge \bigwedge_{i \in I} \varphi_i \Rightarrow \mathit{false}$$
$$(8)\ \forall \overline{x}\, \overline{\alpha} \vee \bigvee_{i \in I} \varphi_i \Rightarrow \mathit{true}$$

Decomposition

$$(9)\ \exists \overline{x}\, \alpha \wedge \bigwedge_{i \in I} \varphi_i\ \Rightarrow (\exists \overline{x}'\, \alpha') \wedge (\forall \overline{x}'\, \overline{\alpha'} \vee \exists \overline{x}''\, \alpha'' \wedge \bigwedge_{i \in I}(\forall \overline{x}'''\, \overline{\alpha'''} \vee \varphi_i))$$
$$(10)\ \forall \overline{x}\, \overline{\alpha} \vee \bigvee_{i \in I} \phi_i \Rightarrow (\forall \overline{x}'\, \overline{\alpha'}) \vee (\exists \overline{x}'\, \alpha' \wedge \forall \overline{x}''\, \overline{\alpha''} \vee \bigvee_{i \in I}(\exists \overline{x}'''\, \alpha''' \wedge \phi_i))$$

Propagation of quantified formulas of A''' + Full elimination (when $I = \emptyset$)

$$(11)\ \exists \overline{x}'''\, \alpha''' \wedge \bigwedge_{i \in I} \overline{\beta_i} \Rightarrow \bigwedge_{i \in I} \forall \overline{x}'''\, \overline{\alpha'''} \vee \overline{\beta_i}$$
$$(12)\ \forall \overline{x}'''\, \overline{\alpha'''} \vee \bigvee_{i \in I} \beta_i \Rightarrow \bigvee_{i \in I} \exists \overline{x}'''\, \alpha''' \wedge \beta_i$$

Direct simplification into A' + Full elimination of quantified formulas of A''

$$(13)\ \exists \overline{x}\, \alpha \wedge \bigwedge_{i \in I} \overline{\beta_i} \Rightarrow \exists \overline{x}'\, \alpha' \wedge \bigwedge_{i \in I'} \overline{\beta_i}$$
$$(14)\ \forall \overline{x}\, \overline{\alpha} \vee \bigvee_{i \in I} \beta_i \Rightarrow \forall \overline{x}'\, \overline{\alpha'} \vee \bigvee_{i \in I'} \beta_i$$

Propagation of false and true

$$(15)\ \exists \overline{x}\, \varphi \wedge \mathit{false} \Rightarrow \mathit{false}$$
$$(16)\ \forall \overline{x}\, \varphi \wedge \mathit{false} \Rightarrow \mathit{false}$$
$$(17)\ \exists \overline{x}\, \varphi \vee \mathit{true} \Rightarrow \mathit{true}$$
$$(18)\ \forall \overline{x}\, \varphi \vee \mathit{true} \Rightarrow \mathit{true}$$

In all these rules I is a finite possibly empty set[2], the φ_i's and the ϕ_i's are working formulas and $\alpha \in AT$.

- In the rules (1),...,(4), the vector $\overline{x}$ is not empty.
- In the rules (5) and (6), the formulas $(\exists x'\, \alpha')$, $(\exists \overline{y}'\, \beta')$ and each $(\exists \overline{z}_i\, \lambda_i)$ belong to A'.
- In the rules (7) and (8), the basic formula $\exists \overline{x}\, \alpha$ is equivalent to a decomposed formula of the form $(\exists \overline{x}'\, \mathit{false} \wedge (\exists \overline{x}''\, \alpha'' \wedge (\exists \overline{x}'''\, \alpha''')))$.
- In the rules (9) and (10), for all $i \in I$, $\varphi_i \notin AT$ and $\overline{\phi_i} \notin AT$. Moreover, the basic formula $\exists \overline{x}\, \alpha$ is equivalent to a decomposed formula of the form $(\exists \overline{x}'\, \alpha' \wedge (\exists \overline{x}''\, \alpha'' \wedge (\exists \overline{x}'''\, \alpha''')))$, with:
  - $\alpha' \neq \mathit{false}$,
  - $\exists \overline{x}'''\, \alpha''' \neq \exists \varepsilon\, \mathit{true}$.
- In the rules (11) and (12):
  - For all $i \in I$, we have $\beta_i \in A'$.
  - $(\exists \overline{x}'''\, \alpha''') \in A'''$.
- In the rules (13) and (14):

---

[2] We recall that if $I = \emptyset$ and φ any first-order constraint then $\bigwedge_{i \in I} \varphi$ is reduced to *true* and $\bigvee_{i \in I} \varphi$ is reduced to *false*.

- The formula $\exists \bar{x}\, \alpha$ is not an element of A' and is equivalent in T to a decomposed formula of the form $\exists \bar{x}'\, \alpha' \wedge (\exists \bar{x}''\, \alpha'' \wedge (\exists \varepsilon\ true))$ with $\alpha' \neq false$.
- For all $i \in I$, we have $\beta_i \in A'$.
- I' is the set of the $i \in I$ such that β_i does not have free occurrences of any variable of $\bar{x}''$.

How does it work? our algorithm follows a clear strategy which decreases the numbers of quantifications until reaching a boolean combination of basic formulas. More precisely, starting from any working formula φ without free variables, the rules $(1),...(4)$ transform φ into a quantified formula of the form $\exists \bar{x}\, \alpha \wedge \bigwedge_{i \in I} \varphi_i$ or $\forall \bar{x}\, \overline{\alpha} \vee \bigvee_{i \in I} \varphi_i$ where φ_i is a working formula. The rules (7) and (8) check that the first level of the quantification is not equivalent to false or true. The rules (9) and (10) decompose then the first level of the quantification and propagate the third part of the decomposition (i.e. formulas which belong to A'''). These steps are repeated until no propagatiopn of A''' can be done. The rules (11), (12) followed by (13) and (14) (all with $I = \emptyset$) eliminate the formulas which belong to A''' and then those which belong to A'' from the deepest formulas. Once these elimination step done only formulas of A' occur in the last level of the formulas. The rules $(11),...,(14)$ can now be applied again with $I \neq \emptyset$ and create formulas of the form $\exists \bar{x}'\, \alpha' \wedge \bigwedge_{i \in I'} \overline{\beta_i'}$ or $\forall \bar{x}'\, \overline{\alpha'} \vee \bigvee_{i \in I'} \beta_i'$. The rules (5) and (6) as well as the other rules can now be applied again and so on. After a finite application of our rules, we get a boolean combination of formulas of the form $\exists \bar{x}'\, \alpha' \wedge \bigwedge_{i \in I'} \overline{\beta_i'}$ or $\forall \bar{x}'\, \overline{\alpha'} \vee \bigvee_{i \in I'} \beta_i'$. Since these formulas have no free variable then from Property 3.2.2 each level is of one of the following forms:

$$(\exists \varepsilon\ true),\ (\exists \varepsilon\ false),\ (\forall \varepsilon\ true),\ (\forall \varepsilon\ false).$$

As a consequence, after a finite application of the rules $(15),...,(18)$ we get either true or false.

Note 1. In order to reach a so simplified final formula, we have also used the following trivial equivalences during the application of our rules:

$(\lambda_1 \wedge \lambda_1)$	$\Rightarrow \lambda_1$	$(\mu_1 \vee \mu_1)$	$\Rightarrow \mu_1$
$(\lambda_1 \wedge true)$	$\Rightarrow \lambda_1$	$(\mu_1 \vee false)$	$\Rightarrow \mu_1$
$(\lambda_1 \wedge \lambda_2)$	$\Rightarrow (\lambda_2 \wedge \lambda_1)$	$(\mu_1 \wedge \mu_2)$	$\Rightarrow (\mu_2 \wedge \mu_1)$
$(\lambda_1 \vee \lambda_2)$	$\Rightarrow (\lambda_2 \vee \lambda_1)$	$(\mu_1 \vee \mu_2)$	$\Rightarrow (\mu_2 \vee \mu_1)$
$(\lambda_1 \wedge (\lambda_2 \wedge \lambda_3))$	$\Rightarrow ((\lambda_1 \wedge \lambda_2) \wedge \lambda_3)$	$(\mu_1 \wedge (\mu_2 \wedge \mu_3))$	$\Rightarrow ((\mu_1 \wedge \mu_2) \wedge \mu_3)$
$(\lambda_1 \vee (\lambda_2 \vee \lambda_3))$	$\Rightarrow ((\lambda_1 \vee \lambda_2) \vee \lambda_3)$	$(\mu_1 \vee (\mu_2 \vee \mu_3))$	$\Rightarrow ((\mu_1 \vee \mu_2) \vee \mu_3)$

where λ_i and μ_i are working formulas.

Correctness of Our Rules

Let us show now that for each rule of the form $p \implies p'$ we have $T \models p \leftrightarrow p'$. The rules $(1),...,(8)$, $(15),...,(18)$ are evident and belong the well known properties of first-order logic. The other rules are new properties of decomposable theories and deserve to be detailed.

Proof of rule (9)

(9) $\exists \bar{x}\, \alpha \wedge \bigwedge_{i \in I} \varphi_i \Rightarrow (\exists \bar{x}'\, \alpha') \wedge (\forall \bar{x}'\, \overline{\alpha'} \vee \exists \bar{x}''\, \alpha'' \wedge \bigwedge_{i \in I}(\forall \bar{x}'''\, \overline{\alpha'''} \vee \varphi_i))$

According to the conditions of application of this rule, the formula $\exists \bar{x}\, \alpha$ is equivalent in T to a decomposed formula of the form $\exists \bar{x}'\, \alpha' \wedge (\exists \bar{x}''\, \alpha'' \wedge (\exists \bar{x}'''\, \alpha'''))$. Thus, the left formula of this rewriting rule is equivalent in T to the formula

$$\exists \bar{x}'\, \alpha' \wedge (\exists \bar{x}''\alpha'' \wedge (\exists \bar{x}'''\alpha''' \wedge \bigwedge_{i \in I} \varphi_i)).$$

Since $\exists \bar{x}'''\, \alpha''' \in A'''$, then according to the fourth point of Definition 2.4.1 we have $T \models \exists! \bar{x}'''\alpha'''$, thus using Property 2.2.3 the preceding formula is equivalent in T to

$$\exists \bar{x}'\, \alpha' \wedge (\exists \bar{x}''\alpha'' \wedge \bigwedge_{i \in I} \neg(\exists \bar{x}'''\alpha''' \wedge \neg\varphi_i))$$

According to the second point of Definition 2.4.1 we have $T \models \exists? \bar{x}'\alpha'$, thus using Property 2.2.2 the preceding formula is equivalent in T to

$$(\exists \bar{x}'\, \alpha') \wedge \neg(\exists \bar{x}'\, \alpha' \wedge \neg(\exists \bar{x}''\alpha'' \wedge \bigwedge_{i \in I} \neg(\exists \bar{x}'''\alpha''' \wedge \neg\varphi_i)))$$

i.e. to

$$(\exists \bar{x}'\, \alpha') \wedge (\forall \bar{x}'\, (\neg\alpha') \vee (\exists \bar{x}''\alpha'' \wedge \bigwedge_{i \in I}(\forall \bar{x}'''(\neg\alpha''') \vee \varphi_i)))$$

which according to Property 3.1.5 is equivalent in T to

$$(\exists \bar{x}'\, \alpha') \wedge (\forall \bar{x}'\, \overline{\alpha'} \vee \exists \bar{x}''\alpha'' \wedge \bigwedge_{i \in I}(\forall \bar{x}'''\overline{\alpha'''} \vee \varphi_i))$$

Proof of rule (10)

(10) $\forall \bar{x}\, \overline{\alpha} \vee \bigvee_{i \in I} \phi_i \Rightarrow (\forall \bar{x}'\, \overline{\alpha'}) \vee (\exists \bar{x}'\, \alpha' \wedge \forall \bar{x}''\, \overline{\alpha''} \vee \bigvee_{i \in I}(\exists \bar{x}'''\, \alpha''' \wedge \phi_i))$

According to the preceding proof we deduce that the left hand side is equivalent in T to the right hand side. Thus, the negation of both hand sides are also equivalent. As a consequence we have:

$$T \models \neg(\exists \bar{x}\alpha \wedge \bigwedge_{i \in I} \varphi_i) \leftrightarrow \neg((\exists \bar{x}'\, \alpha') \wedge (\forall \bar{x}'\, \overline{\alpha'} \vee \exists \bar{x}''\, \alpha'' \wedge \bigwedge_{i \in I}(\forall \bar{x}'''\, \overline{\alpha'''} \vee \varphi_i)))$$

By pushing down the negation from both sides and according to Property 3.1.5 we get

$$T \models \forall \bar{x}\, \overline{\alpha} \vee \bigvee_{i \in I} \phi_i \leftrightarrow (\forall \bar{x}'\, \overline{\alpha'}) \vee (\exists \bar{x}'\, \alpha' \wedge \forall \bar{x}''\, \overline{\alpha''} \vee \bigvee_{i \in I}(\exists \bar{x}'''\, \alpha''' \wedge \phi_i))$$

where ϕ_i is the formula $\neg\varphi_i$. According to the condition of application of rule (9) we have $\varphi_i \notin AT$ then according to Property 3.1.5 we get $\overline{\phi_i} \notin AT$.

Proof of rule (11)

$$(11) \quad \exists \overline{x}''' \, \alpha''' \wedge \bigwedge_{i \in I} \overline{\beta_i} \Rightarrow \bigwedge_{i \in I} \forall \overline{x}''' \overline{\alpha'''} \vee \overline{\beta_i}$$

According to Property 3.1.5, the left hand side of rule (11) is equivalent in T to

$$\left(\exists \overline{x}''' \, \alpha''' \wedge \bigwedge_{i \in I} \neg \beta_i \right)$$

Since $\exists \overline{x}''' \, \alpha''' \in A'''$, then according to the fourth point of Definition 2.4.1 we have $T \models \exists ! \overline{x}''' \alpha'''$, thus according to Property 2.2.3 the preceding formula is equivalent in T to

$$\bigwedge_{i \in I} \neg (\exists \overline{x}''' \, \alpha''' \wedge \beta_i)$$

i.e. to

$$\bigwedge_{i \in I} (\forall \overline{x}''' \, \neg \alpha''' \vee \neg \beta_i)$$

which according to Property 3.1.5 is equivalent to

$$\bigwedge_{i \in I} (\forall \overline{x}''' \, \overline{\alpha'''} \vee \overline{\beta_i})$$

Proof of rule (12)

$$(12) \quad \forall \overline{x}''' \, \overline{\alpha'''} \vee \bigvee_{i \in I} \beta_i \Rightarrow \bigvee_{i \in I} \exists \overline{x}''' \alpha''' \wedge \beta_i$$

According to the preceding proof we deduce that the left hand side is equivalent in T to the right hand side. Thus, the negation of both hand sides are also equivalent. As a consequence we have:

$$T \models \neg (\exists \overline{x}''' \, \alpha''' \wedge \bigwedge_{i \in I} \overline{\beta_i}) \leftrightarrow \neg (\bigwedge_{i \in I} \forall \overline{x}''' \overline{\alpha'''} \vee \overline{\beta_i})$$

By pushing down the negation from both sides and according to Property 3.1.5 we get

$$T \models \forall \overline{x}''' \, \overline{\alpha'''} \vee \bigvee_{i \in I} \beta_i \leftrightarrow \bigvee_{i \in I} \exists \overline{x}''' \alpha''' \wedge \beta_i$$

Proof of rule (13)

$$(13) \quad \exists \overline{x} \, \alpha \wedge \bigwedge_{i \in I} \overline{\beta_i} \Rightarrow \exists \overline{x}' \, \alpha' \wedge \bigwedge_{i \in I'} \overline{\beta_i}$$

- the formula $\exists \overline{x} \, \alpha$ is not an element of A' and is equivalent in T to a decomposed formula of the form $\exists \overline{x}' \, \alpha' \wedge (\exists \overline{x}'' \, \alpha'' \wedge (\exists \varepsilon \, true))$ with $\alpha' \neq false$.
- For all $i \in I$, we have $\beta_i \in A'$.

- I' is the set of the $i \in I$ such that β_i does not have free occurrences of any variable of $\bar{x}''$.

According to the preceding conditions, the left hand side of rule (13) is equivalent in T to

$$\exists \bar{x}' \, \alpha' \wedge (\exists \bar{x}'' \, \alpha'' \wedge \bigwedge_{i \in I} \overline{\beta_i})$$

Let us denote by I_1, the set of the $i \in I$ such that x_n'' does not have free occurrences in the formula β_i, thus the preceding formula is equivalent in T to

$$\exists \bar{x}' \, \alpha' \wedge (\exists x_1''...\exists x_{n-1}'' \left[\begin{matrix} (\bigwedge_{i \in I_1} \overline{\beta_i}) \wedge \\ (\exists x_n'' \, \alpha'' \wedge \bigwedge_{i \in I - I_1} \overline{\beta_i}) \end{matrix} \right]). \tag{8}$$

Since $\exists \bar{x}'' \alpha'' \in A''$ and $\beta_i \in A'$ for every $i \in I - I_1$, then according to Property 2.3.2 and the conditions 2 and 3 of Definition 2.4.1, the formula (8) is equivalent in T to

$$\exists \bar{x}' \, \alpha' \wedge (\exists x_1''...\exists x_{n-1}'' ((\bigwedge_{i \in I_1} \overline{\beta_i}) \wedge true)). \tag{9}$$

By repeating the three preceding steps $(n-1)$ times, by denoting by I_k the set of the $i \in I_{k-1}$ such that $x_{(n-k+1)}''$ does not have free occurrences in β_i, and by using $(n-1)$ times Property 2.3.3, the preceding formula is equivalent in T to

$$\exists \bar{x}' \, \alpha' \wedge \bigwedge_{i \in I_n} \overline{\beta_i}$$

Proof of rule (14)

$$(14) \quad \forall \bar{x} \, \overline{\alpha} \vee \bigvee_{i \in I} \beta_i \Rightarrow \forall \bar{x}' \, \overline{\alpha}' \vee \bigvee_{i \in I'} \beta_i$$

According to the preceding proof we deduce that the left hand side is equivalent in T to the right hand side. Thus, the negation of both hand sides are also equivalent. As a consequence we have:

$$T \models \neg(\exists \bar{x} \, \alpha \wedge \bigwedge_{i \in I} \overline{\beta_i}) \Rightarrow \neg(\exists \bar{x}' \, \alpha' \wedge \bigwedge_{i \in I'} \overline{\beta_i})$$

By pushing down the negation from both sides and according to Property 3.1.5 we get

$$T \models \forall \bar{x} \, \overline{\alpha} \vee \bigvee_{i \in I} \beta_i \leftrightarrow \forall \bar{x}' \, \overline{\alpha}' \vee \bigvee_{i \in I'} \beta_i$$

Example 3.3.1. *Let us solve the following formula φ_1 in the theory of finite or infinite trees:*

$$\exists x \forall y \, \exists z \, z = f(y) \wedge x = f(x) \vee (x = f(y) \wedge z = f(z))$$

According to the rule (3) the preceding formula is equivalent to

$$\exists x \forall y \, (\exists z \, z = f(y) \wedge x = f(x)) \vee (\exists z \, x = f(y) \wedge z = f(z)).$$

By applying the rule (9) with $I = \emptyset$ on $(\exists z\, z = f(y) \wedge x = f(x))$ and also on $(x = f(y) \wedge z = f(z))$, we get the following equivalent formula

$$\exists x \forall y\, ((x = f(x)) \wedge (x = f(x))) \vee ((x = f(y)) \wedge ((x = f(y))),$$

In fact:

- *the formula $(\exists z\, z = f(y) \wedge x = f(x))$ is equivalent to a decomposed formula of the form $(\exists \varepsilon\, x = f(x) \wedge (\exists \varepsilon\, true \wedge (\exists z\, z = f(y))))$.*
- *the formula $(\exists z\, x = f(y) \wedge z = f(z))$ is equivalent to a decomposed formula of the form $(\exists \varepsilon\, x = f(y) \wedge (\exists \varepsilon\, true \wedge (\exists z\, z = f(z))))$.*
- *We also recall that if $I = \emptyset$ then $\bigwedge_{i \in I} \varphi_i$ is the formula true and $\bigvee_{i \in I} \varphi_i$ is the formula false.*

According to Note 1, the preceding formula is simplified into

$$\exists x \forall y\, x = f(x) \vee x = f(y).$$

Since $\exists x\, true \in A''$, then rule (13) can be applied. The preceding formula is equivalent to the empty conjunction, i.e. the formula true.

4   Benchmarks: Randomly Generated Formulas

We have tested our rules on randomly generated formulas of the form

$$\exists \bar{x}_1\, \alpha_1 \wedge \forall \bar{y}_1\, \beta_1 \wedge \exists \bar{x}_2\, \alpha_2 \wedge \forall \bar{y}_2\, \beta_2 ... \tag{10}$$

such that in each sub-formula of the form $\exists \bar{x}_i\, \alpha_i \wedge \forall \bar{y}_i\, \beta_i$ we have:

- Each $\bar{x}_i$ or $\bar{y}_i$ is a vector of variables whose cardinality is randomly chosen between 1 and 2.
- The formulas $\exists \bar{x}_i \alpha_i$ and $\forall \bar{y}_i \beta_i$ are basic formulas.
- The number of the formulas in the α_i and β_i is randomly chosen between 1 and 5. Moreover, the formulas *true* or *false* occur at most once.
- The formulas are randomly generated starting from a set containing two 1-ary function symbols: $f, -$ and two 2-ary function symbols $+, g$.

The benchmarks are realized on two decomposable theories:

(1) theory T_1 of finite or infinite trees [6],

(2) theory T_2 of the combination of finite or infinite trees and rational numbers with addition and subtraction [9]. It represents an axiomatization of the model of Prolog III [4] that we presented in [8].

We have used a 2.5Ghz Pentium IV processor with 1024Mb of RAM. For each integer $1 \leq AQ \leq 16$ (AQ stands for alternated quantifications $\exists \forall$) we randomly generated formulas with AQ nested alternated quantifications of the form $\exists \bar{x}_i\, \alpha_i \wedge \forall \bar{y}_i\, \beta_i$, we solved them and computed the average execution time (CPU time in milliseconds). If for example $AQ = 2$ then we solved formulas of the form $\exists \bar{x}_1\, \alpha_1 \wedge \forall \bar{y}_1\, \beta_1 \wedge \exists \bar{x}_2\, \alpha_2 \wedge \forall \bar{y}_2\, \beta_2$.

For each theory T_i, we note bellow the average execution times obtained using our rules as well as those obtained using the classical decision procedure for decomposable theories [7].

Theory T_1.

AQ	2	4	6	10	15	16
Algo [7] T_1	132	490	1592	19234	–	–
Our algo T_1	92	231	704	9252	2556624	–

Theory T_2.

AQ	2	4	6	9	13	14
Algo [7] T_2	206	682	2188	69234	–	–
Our algo T_2	128	342	1128	32040	3011028	–

For both theories, the classical decision procedure takes much more time and overflows the memory early (starting from AQ=10 for T_1 and 9 for T_2) comparing with our new algorithm which can decide the truth value of formulas having more than 15 alternations of $\forall\exists$ (i.e. 30 quantifiers) in T_1 and 13 (i.e. 26 quantifiers) in T_2.

This big differences (in time and space) is due to the fact that the decision procedure of [7] needs to transform the initial constraint (10) into a normalized formula before starting solving it. Such a transformation creates a complex and huge normalized formula (in term of depth) with many imbricated alternated quantifiers. This later will be solved mainly using a very costly rule (rule (5) in [7]) which decreases the depth of the normalized formulas but increases exponentially the number of conjunctions of the normalized formulas until overflowing the memory. Our algorithm does not need to normalize the initial constraint (10) and creates instead a boolean combination of basic formulas until reaching the solved constraint *true* or *false*.

5   Conclusion

We have presented in this paper an efficient decision procedure for functional decomposable theories, i.e. theories whose set of relation is reduced to $\{=, \neq\}$ such as: Clark equational theory [2], theory of finite or infinite trees [11], rational or real numbers with addition and subtraction, theory of queues [14] together with the two functions add-left and add-right,...etc. To this end, we presented the notion of dual and working formulas and used them to show how to transform any first-order proposition into a boolean combination of formulas which are either equivalent to true or to false.

The classical decision procedure for decomposable theories given in [7] needs to normalize the initial proposition before starting the solving process. This transformation generates many nested negations and quantifications which greatly slow down the performances of the solver. Our new algorithm does not need such a transformation and directly computes the truth value of any proposition using new properties of functional decomposable theories.

We have shown the efficiency of our algorithm (in time and space) by comparing its performances with those of the classical decision procedure for decomposable theories [7]. Our algorithm can solve propositions involving more than 30 nested alternated quantifiers ($\exists\forall$) while the decision procedure overflows the memory starting from 20 nested alternated quantifiers.

Recently, we transformed the classical decision procedure into a full first-order constraint solver which can manage free variables and present their solutions in a clear and explicit way [5]. After many experimentations, we noted that the performances of this later are still worse (in time and space) comparing with our dual constraint algorithm while solving formulas of the form $\exists \bar{x}_1\, \alpha_1 \wedge \forall \bar{y}_1\, \beta_1 \wedge \exists \bar{x}_2\, \alpha_2 \wedge \forall \bar{y}_2\, \beta_2$. This is mainly due to the fact that we kept our pre-processing step which transforms the initial constraint into a normalized formula before starting the solving process. The reason for keeping this step is simple:

(1) If we remove the pre-processing step then we can only create a boolean combination of basic formulas. In this case, if there exists at least one free variable then we can get some basic formulas with free variables which are neither equivalent to false nor to true. As a consequence, our final formula will be a boolean combination from which it is impossible to understand the values of the free variables which satisfy the formula in all the models of T.

(2) If we keep the pre-processing step then we handle only normalized formulas during the solving process. As a consequence, it is very easy to extract the solutions of the free variables in the final solved form.

From (1) and (2) we can deduce that if we want a full first-order constraint solver which manages free variables than we cannot avoid the pre-processing step which normalizes the initial constraint. The dual algorithm given in this paper is not able to manage free variables but is by far the most efficient when deciding the truth values of any proposition, i.e. formula without free variables.

Currently, we are trying to find a more abstract characterization and/or a model theoretic characterization of the decomposable theories. The current definition gives only an algorithmic insight into what it means for a theory to be complete.

Our work as presented in this paper contributes to enlarge the properties of decomposable theories from a theoretical point of view but does not really show the efficiency of our work on real practical problems. As a consequence, we are now working on more meaningful (i.e., non random) benchmarks and try to combine our approach with those of Satisfiability Modulo Theories (SMT) community [15]. Many theories are of interests in particular: theory of lists [16] and Presburger's theory [12,13].

References

1. Apt, K.: Principles of constraint programming. Cambridge University Press, Cambridge (2003)
2. Clark, K.L.: Negation as failure. In: Gallaire, H., Minker, J. (eds.) Logic and Data bases. Plenum Pub., New York (1978)
3. Colmerauer, A., Dao, T.: Expressiveness of full first-order constraints in the algebra of finite or infinite trees. Journal of Constraints 8(3), 283–302 (2003)
4. Colmerauer, A.: An introduction to Prolog III. Communication of the ACM 33(7), 68–90 (1990)
5. Djelloul, K.: A full first-order constraint solver for decomposable theories. In: Autexier, S., Campbell, J., Rubio, J., Sorge, V., Suzuki, M., Wiedijk, F. (eds.) AISC 2008, Calculemus 2008, and MKM 2008. LNCS (LNAI), vol. 5144, pp. 93–108. Springer, Heidelberg (2008)

6. Djelloul, K., Dao, T., Fruehwirth, T.: Theory of finite or infinite trees revisited. Theory and practice of logic programming (TPLP) 8(4), 431–489 (2008)
7. Djelloul, K.: Decomposable theories. Theory and practice of logic programming (TPLP) 7(5), 583–632 (2007)
8. Djelloul, K., Dao, T.: Extension into trees of first-order theories. In: Calmet, J., Ida, T., Wang, D. (eds.) AISC 2006. LNCS (LNAI), vol. 4120, pp. 53–67. Springer, Heidelberg (2006)
9. Djelloul, K.: About the combination of trees and rational numbers in a complete first-order theory. In: Gramlich, B. (ed.) FroCos 2005. LNCS (LNAI), vol. 3717, pp. 106–121. Springer, Heidelberg (2005)
10. Fruehwirth, T., Abdennadher, S.: Essentials of Constraint Programming. Springer, Heidelberg (2003)
11. Maher, M.: Complete Axiomatizations of the Algebras of Finite, Rational and Infinite Trees. In: Proc. of LICS 1988 Annual Symposium on Logic in Computer Science, pp. 348–357 (1988)
12. Oppen, D.: A $2^{2^{2^n}}$ Upper Bound on the Complexity of Presburger Arithmetic. Journal of Comput. Syst. Sci. 16(3), 323–332 (1978)
13. Presburger, M.: Uber die Vollstandigkeit eines gewissen Systems der Arithmetik ganzer Zahlen, in welchem die Addition als einzige Operation hervortritt. In: Comptes Rendus du I congrs de Mathematiciens des Pays Slaves, Warszawa, pp. 92–101 (1929)
14. Rybina, T., Voronkov, A.: A decision procedure for term algebras with queues. ACM transaction on computational logic 2(2), 155–181 (2001)
15. Satisfiability Modulo Theories (SMT) web page,
   `http://combination.cs.uiowa.edu/smtlib/`
16. Spivey, J.: A Categorial Approch to the Theory of Lists. In: van de Snepscheut, J.L.A. (ed.) MPC 1989. LNCS, vol. 375, pp. 399–408. Springer, Heidelberg (1989)
17. Vorobyov, S.: An improved lower bound for the elementary theories of trees. In: McRobbie, M.A., Slaney, J.K. (eds.) CADE 1996. LNCS, vol. 1104, pp. 275–287. Springer, Heidelberg (1996)

Challenges in Constraint-Based Analysis of Hybrid Systems*

Andreas Eggers[1], Natalia Kalinnik[2], Stefan Kupferschmid[2], and Tino Teige[1]

[1] Carl von Ossietzky Universität Oldenburg, Germany
{eggers,teige}@informatik.uni-oldenburg.de
[2] Albert-Ludwigs-Universität Freiburg, Germany
{kalinnik,skupfers}@informatik.uni-freiburg.de

Abstract. In the analysis of hybrid discrete-continuous systems, rich arithmetic constraint formulae with complex Boolean structure arise naturally. The iSAT algorithm, a solver for such formulae, is aimed at bounded model checking of hybrid systems. In this paper, we identify challenges emerging from planned and ongoing work to enhance the iSAT algorithm. First, we propose an extension of iSAT to directly handle ordinary differential equations as constraints. Second, we outline the recently introduced generalization of the iSAT algorithm to deal with probabilistic hybrid systems and some open research issues in that context. Third, we present ideas on how to move from bounded to unbounded model checking by using the concept of interpolation. Finally, we discuss the adaption of some parallelization techniques to the iSAT case, which will hopefully lead to performance gains in the future. By presenting these open research questions, this paper aims at fostering discussions on these extensions of constraint solving.

Keywords: mixed Boolean and arithmetic constraints, differential equations, stochastic SMT, Craig interpolation, parallel solver.

1   Introduction

The complexity of embedded systems, e.g. in automotive and avionics applications, has increased dramatically over the last decades. The safety criticality of these systems calls for more and more sophisticated — especially computer-aided — analysis techniques that enable engineers to assess the correctness of their designs and implementations. For finding errors in models of large systems, simulation has become one of the most successful and established methods. However, in general, simulation cannot guarantee the absence of errors for systems with infinitely many states which naturally arise in these domains.

---

* This work has been partially supported by the German Research Council (DFG) as part of the Transregional Collaborative Research Center "Automatic Verification and Analysis of Complex Systems" (SFB/TR 14 AVACS, www.avacs.org).

A. Oddi, F. Fages, and F. Rossi (Eds.): CSCLP 2008, LNAI 5655, pp. 51–65, 2009.
© Springer-Verlag Berlin Heidelberg 2009

In recent years, algorithms have been developed that can mathematically prove the correctness of a huge variety of system classes with respect to a given specification. Embedded systems often combine digital and analog components, e.g. in multi-modal controllers or when describing them as integrated models of a digital controller interacting with its continuously evolving plant. *Hybrid systems* are a very rich modelling paradigm to describe such hybrid discrete-continuous behavior. A hybrid system consists of a set of modes and a set of continuous variables that together represent its state space. Its evolution is described by a transition relation entailing discrete mode switches, also called transitions, and arithmetic constraints describing the behavior of the continuous variables within each mode. The latter is often achieved by using differential equations that naturally arise when modelling physical entities. The mode switches are governed by so-called transition guards, i.e. arithmetic constraints observing the continuous variables, and can perform discrete actions by (potentially non-deterministically) setting a variable x to a new value x' satisfying an arithmetic condition, e.g. $x' > \sin(y^2)$ or $x' = 4.2 \cdot x$.

The semantics of a hybrid system is defined by the set of its runs, i.e. the possible evolutions it allows. Such an evolution can always be represented by a sequence of variable valuations, where two successive valuations can either be connected by a continuous evolution in the mode the system is in, or satisfy the transition guard and action constraints, such that the system can actually perform a switch from one mode to the next. This representation of a run is called a trace and intuitively describes snapshots of the system's evolution through the state space at the endpoints of continuous trajectories and discrete jumps.

Returning to the motivation described initially, the reachability problem of hybrid systems, i.e. the question of whether a particular state (e.g. a state representing a fatal system failure) is reachable, is of particular interest to complex systems verification and falsification. Though this problem is undecidable in general, developing model checking algorithms and tools that can deal with a large sample of systems that occur in real-world applications seems to be so relevant that it can be considered a reasonable goal nonetheless. In addition to that, robustness notions [1] can be used to find classes of systems, for which decision procedures can be developed. Hybrid systems and decidability questions have been extensively examined in the literature, for a detailed account see e.g. [2].

Among the most successful analysis methods for finite-state systems is *bounded model checking* (BMC) [3,4], which has also been extended to the case of hybrid systems [5,6]. The idea of BMC is to encode the initial states, the transition relation, and the target state specification of the system as predicates $INIT(\boldsymbol{x}_0)$, $TRANS(\boldsymbol{x}_i, \boldsymbol{x}_{i+1})$, and $TARGET(\boldsymbol{x}_k)$, respectively, where $\boldsymbol{x}_0, \boldsymbol{x}_i, \boldsymbol{x}_{i+1}$, and $\boldsymbol{x}_k$ are instantiations of the vector of variables representing the discrete and continuous state space. The initial predicate $INIT(\boldsymbol{x}_0)$ is satisfied by a valuation of $\boldsymbol{x}_0$ iff that valuation characterizes an initial state. Analogously, the transition predicate $TRANS(\boldsymbol{x}_i, \boldsymbol{x}_{i+1})$ holds for two (successive) valuations iff the system can perform a discrete mode switch or a continuous evolution as described above. We consider the succession from $\boldsymbol{x}_i$ to $\boldsymbol{x}_{i+1}$ as a *step* of the system. Finally, the target predicate $TARGET(\boldsymbol{x}_k)$ specifies the states whose reachability is examined. A hybrid system can thus reach a target state in a limited number of steps k iff the following BMC formula is satisfiable.

$$\Phi_k := INIT(\boldsymbol{x}_0) \wedge \underbrace{TRANS(\boldsymbol{x}_0, \boldsymbol{x}_1) \wedge \ldots \wedge TRANS(\boldsymbol{x}_{k-1}, \boldsymbol{x}_k)}_{k \text{ unwindings of the transition relation}} \wedge TARGET(\boldsymbol{x}_k)$$

As the behavior of a hybrid system can in general be arbitrarily non-linear and non-deterministic, the resulting BMC formula Φ_k is a Boolean combination of rich arithmetic constraints including differential equations. A solver that can directly handle Φ_k is thus desirable. Approaches from continuous constraint programming (cf. e.g. [7]) which can handle non-linear constraints are often restricted to conjunctive formulae. On the other hand, most *satisfiability modulo theories* (SMT, e.g. [8]) solvers — though being very capable of handling complex Boolean structure — are confined to decidable theories — in particular, they do not handle non-linear constraints. Recently, algorithms combining both domains were published: ABSOLVER [9], which uses a non-linear optimization packet, and iSAT [10], which uses techniques from interval constraint solving.

Structure of the paper. In Section 2, we briefly recall the iSAT algorithm that constitutes the basic framework for our presentation. Section 3 identifies the main directions for the extensions discussed in this paper, which are described in more detail in Sections 4–7. Finally, we give some thoughts on synergies between these different topics.

2   The iSAT Algorithm

In [10], the iSAT algorithm for solving mixed Boolean and non-linear (including transcendental) arithmetic constraints over bounded reals and integers was introduced. Differential equations however cannot be handled directly by iSAT and need to be (over-)approximated or solved during modeling. Internally, iSAT solves a conjunction of clauses, where a clause is a disjunction of atoms. An atom (a.k.a. primitive constraint) contains one relational operator, at most one arithmetic operation, and up to three variables, e.g. $x \geq \sin(y)$, $x = y+z$, and $z < 3.7$. By a Tseitin-like transformation [11], any BMC formula Φ_k can automatically be rewritten into an equi-satisfiable formula in this kind of conjunctive normal form, which grows at most linearly in the size of the original formula. The iSAT algorithm is a generalization of the Davis-Putnam-Logemann-Loveland (DPLL) procedure [12,13] using interval constraint propagation (ICP) (cf. e.g. [7]), and manipulates interval valuations of the variables by alternating *deduction* and *splitting* phases.

During the *deduction* phase, the solver searches for clauses in which all but one atom are inconsistent under the current interval valuation. These remaining consistent atoms are called *unit*. In order to retain a chance for satisfiability of the formula, unit atoms have to be satisfied. This is similar to Boolean constraint propagation in DPLL SAT solving. The unit atoms are therefore used for ICP during the deduction phase. New interval bounds can thus be deduced until a fixed point is reached. For termination reasons, the ICP has to be stopped if the progress of newly deduced bounds becomes negligible.

If a *conflict* occurs, i.e. the interval of a variable becomes empty, then a conflict resolution procedure is called which analyzes the reason for the conflict. If the conflict cannot be resolved the given formula is unsatisfiable. Otherwise, a conflict clause is built from the reason of the conflict and added to the formula in order to prevent the solver from revisiting the same conflict again. Furthermore, conflict resolution requires the solver to take back some of the decisions and their accompanying deductions that have been performed so far.

If the solver finds a *solution*, i.e. at least one atom in each clause is satisfied by every point in the interval valuation, the algorithm stops. In general, equations like $x = y \cdot z$ can only be satisfied by point intervals. However, reaching such point intervals by ICP cannot be guaranteed for continuous domains. One option to mitigate this problem is to stop the search when all intervals have a width smaller than a certain threshold, the so-called *minimum splitting width*, and returning the found *approximative* solution. Having completed the deduction phase and neither found a conflict nor an (approximative) solution, iSAT performs a decision by *splitting* an interval. A decision heuristics is used to select one of the intervals whose width is still greater than or equal to the minimum splitting width. The search is then resumed using this newly generated interval bound which potentially triggers new deductions as described above.

3   Problem Description

The primary goal of this paper is to propose challenges that arise from planned and ongoing work in the context of enhancing the iSAT algorithm into the directions of scope and performance. We hope that presenting these research questions will foster discussions on these interesting topics.

In order to avoid an a priori overapproximation of the continuous dynamics in system models, direct handling of ordinary differential equations is to be integrated into iSAT's deduction rules (Section 4). Another extension of the scope is to enable iSAT to deal with probabilistic hybrid systems (Section 5). Thereafter, we present ideas on how to move from bounded to unbounded model checking by using the concept of interpolation. In Section 7, we discuss the adaption of some parallelization techniques to the iSAT case, which will hopefully lead to dramatic performance gains in the future.

4   Differential Equations as Constraints

In order to directly handle ordinary differential equations (ODEs) as constraints within a formula, we need to extend the deduction mechanism used by iSAT to not only propagate newly deduced bounds through arithmetic constraints using ICP, but to also propagate new interval bounds through ODEs. A continuous trajectory can often be described by an ODE of the form $\frac{d\boldsymbol{x}}{dt} = f(\boldsymbol{x})$ over a vector x of continuous variables. Being part of the predicative encoding of the transition relation, this ODE describes the connection of the variable instances x_i and x_{i+1} from two successive BMC unwinding depths. We thus search for solution functions of the ODE that emerge from the valuation of x_i (the *pre-box*) and eventually reach the valuation of x_{i+1} (the *postbox*). Analogously to ICP, we are interested in pruning off all valuations from the pre- and postbox that cannot belong to such trajectories. To achieve this, we thus need a safe overapproximation of the ODE trajectories in order not to prune away possible solutions.

Related work on safe enclosures of ODEs can be found in [14,15,16], where error bounds on the remainder term of a *Taylor series* of the unknown exact solution are calculated and used as safe overapproximations of the errors induced by the Taylor-series-based approximation of the trajectory. Using coordinate transformations to suitably adapt the enclosures to the solution sets and thereby mostly avoiding the so-called *wrapping effect* (cf. e.g. [14]), this approach works well for linear ODEs. However, as for non-linear ODEs, coordinate transformations alone are often insufficient to eliminate the wrapping effect, the enclosures quickly become very rough and finally unusably large in the non-linear case. A newer approach — the so-called *Taylor models* [17] — have been shown to give tighter enclosures for non-linear ODEs by employing a more symbolic representation of the enclosure sets. Henzinger et. al. use the Taylor-series-based enclosure method in the HYPERTECH tool [18], also facing hybrid systems analysis, however not in a constraint solving approach. In [19], CLP(F) — a very broad framework to constraint logic programming — is applied to models of hybrid systems. CLP(F) does however not include any measures against the wrapping effect encountered when enclosing ODE trajectories with intervals.

This section of the paper tries to sum up the essential challenges and options to solve them, that we see in the context of embedding safe enclosures of ODEs into the iSAT algorithm. These major challenges are to

1. find methods and data structures that allow sufficiently tight overapproximating enclosures of the trajectories of an ODE that connect the interval boxes representing pre- and postsets,
2. devise heuristics that allow to select the method fitting best into the current phase of solving, e.g. coarse-grain but quick first enclosures to chop off the most implausible parts of the solution space during an early phase of solving versus tight but expensive enclosures to narrow an enclosure tube around an actual error trace to reduce the probability of spurious counterexamples,
3. embed these methods in the solver process anywhere between calling them like normal deduction rules that are executed whenever a new bound on a variable is generated or as a subordinate solver that can be called arbitrarily seldom to reduce the impact of an expensive method,
4. derive symbolic knowledge from ODE constraints that can be learned and thus automatically extend the constraint system to contain redundant encodings reflecting some of the possible system dynamics without the need to (probably always more expensively) re-enclose the ODE trajectories.

In a first prototype, we have proved the feasibility of integrating a Taylor-series based enclosure method as a subordinate solver to the iSAT algorithm. However, first experiments with this prototype show that none of above challenges can be regarded as completely mastered [20]. For each challenge a multitude of design choices exist that may have a fundamental impact on the overall performance of an ODE-enabled iSAT.

To approach the first challenge, we consider both the Taylor-based enclosure methods [14,15,16], that were used for the prototype, and the more recently developed Taylor models [17] as possible choices. While Taylor series together with coordinate transformations will probably be a good choice for linear ODEs, we expect Taylor models to also allow to approach non-linear ODEs. Out of the many existing numerical methods for numerical approximation of ODE trajectories there may however be some whose truncation errors can be enclosed as well. Exploring such methods may thus extend the spectrum of choices. The most essential problem in this context will probably be to control the wrapping effect in order to avoid unusably coarse overapproximations.

The second challenge necessitates, first, a pool of methods with different characteristics, i.e. methods that are tailored to quickly generating results as well as methods that allow tight enclosures, and second, a set of criteria that are easily evaluable and allow to determine which enclosure method should be used. One possible criterion could be the size of the currently searched box, where a small box could indicate the use of more accurate methods.

Solving the third challenge will mean to find the best integration strategy for the enclosure mechanisms. It can be expected that any good enclosure method will normally be quite expensive in terms of runtime compared to arithmetic interval propagators. This may suggest that performing enclosures very seldom might be a good strategy. On the other hand, as an essential portion of the system dynamics is encoded in the ODE constraints, it also seems necessary to evaluate them often in order to detect conflicts early and thus to prune off those parts of the search space that cannot contain any solutions.

Finally, we expect that learning new arithmetic constraints from the ODEs (e.g. from monotony or stability arguments) will allow to reduce the number of enclosures that actually need to be performed. Similar to learning conflict clauses when the intersection of an enclosure with a pre- or postbox becomes empty, these constraints would allow to prune off substantial parts of the search space that cannot contain any solutions.

5   Stochastic Constraint Systems

Most of the common analysis procedures are confined to just prove or disprove the safety of a system. However, for models of safety-critical systems interacting with the environment it is often clear which incidents lead to unsafe behavior, e.g. a power blackout combined with a failure of the emergency power system can induce a safety-critical state of a nuclear power station. Although such accidents cannot be excluded in general, it is tried to strongly decrease the probability of safety-critical behavior s.t. the system is *most likely* safe. Thus, the

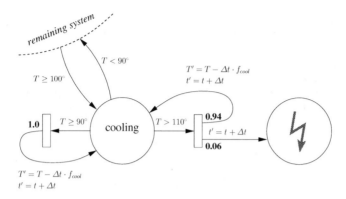

Fig. 1. A fragment of a system model for a probabilistic component breakdown

verification goal in this application domain is whether the *likelihood* of reaching unsafe states is below an acceptable threshold, e.g. less than 0.03%. As a modeling framework to deal with uncertainties, we consider *probabilistic hybrid automata* (PHA, cf. [21]) which extends the notion of hybrid automata s.t. the non-deterministic selection of a transition is enriched by a probabilistic choice according to a distribution over variants of the transition. I.e., each transition carries a (discrete) probabilistic distribution. Each probabilistic choice within such a distribution leads to a potentially different successor mode while performing some discrete actions. An example of a PHA fragment modelling some probabilistic component breakdown is shown in Fig. 1, where T, f_{cool}, and t denote the temperature, the cooling factor, and the time, resp., and Δt is the discretization parameter.

In order to automatically compute the reachability probability of (un-)desired properties of PHAs, in [21] we introduced *stochastic satisfiability modulo theories* (SSMT) as the unification of stochastic propositional satisfiability (SSAT) [22] and satisfiability modulo theories (SMT, e.g. [8]). SSMT deals with *existential* and *randomized* quantification of finite-domain variables. An SSMT formula is specified by a quantifier prefix and an SMT formula, e.g. $\exists x \in \{0,1\}$ $\text{\fontfamily{cmr}\selectfont R}_{\langle(0,0.6),(1,0.4)\rangle}y \in \{0,1\} : (x > 0 \vee \sin(4a) \geq 0.3) \wedge (y > 0 \vee \sin(4a) < 0)$. The value of a variable bound by an existential quantifier, as in $\exists x \in \{0,1\}$, can be set arbitrarily, while the value of a variable bound by a randomized quantifier, as in $\text{R}_{\langle(0,0.6),(1,0.4)\rangle}y \in \{0,1\}$, is determined stochastically by the corresponding distribution, here $\langle(0,0.6),(1,0.4)\rangle$. E.g., $\text{R}_{\langle(0,0.6),(1,0.4)\rangle}y \in \{0,1\}$ means that the variable y is assigned the value 0 or 1 with probability 0.6 or 0.4, resp. The solution of an SSMT problem Φ is a tree of assignments to the existential variables that *maximizes the overall satisfaction probability* of Φ (cf. [21] for more details). In our application, we are interested in computing the maximum probability of satisfaction. For the example above, setting x to 0 yields the satisfaction probability 0.4 since the assignment $x = 0, y = 0$ cannot satisfy the SMT formula. For $x = 1$, both $y = 0$ and $y = 1$ lead to solutions and, thus, to satisfaction probability 1.0. Hence, the maximum satisfaction probability is 1.0.

The behavior of a PHA $\mathcal{H}$ (restricted to step depth k) together with a reachability property P can be described by an SSMT formula Φ in the following sense: Φ is satisfiable with maximum probability p iff $\mathcal{H}$ fulfills property P (within k steps) with maximum probability p. The idea of the formalized encoding of a PHA into an SSMT formula, as presented in [21], is that the non-deterministic choice of a transition in a PHA corresponds to existential quantification in SSMT, while the probabilistic distributions correspond to randomized quantification. The discrete-continuous behavior of the automaton is encoded by means of standard techniques. We are currently working on a modeling framework for PHAs which automatically translates the modelled PHA into an SSMT formula [23].

Completing the verification procedure for PHAs, we recently extended the iSAT solver to existential and randomized quantification (SiSAT, [24]). The main idea of the SiSAT algorithm is to traverse the tree given by the quantifier prefix and to properly call the iSAT algorithm. First experimental results proved the concept: exploiting additional pruning rules which cut off huge parts of the quantifier tree, the SiSAT tool is currently able to solve SSMT problems with up to 110 quantified and 350 non-quantified variables, and up to 1100 clauses within 100 minutes. Problems including quantifiers are well-known not to be as scalable as quantifier-free problems. However, we believe that further improvements, e.g. value and variable ordering heuristics, will yield significant performance gains. In the following, we highlight some open research issues for future work.

- *Value and variable ordering heuristics* are well-studied in SAT and Constraint Satisfaction to improve efficiency. For the quantified case, the variable ordering within a block of quantifiers with the *same* type do not change the semantics of the problem. This property can be exploited during the proof search to rearrange the variables. To derive benefit from this fact, we will investigate different static and dynamic ordering heuristics.
- *Bounded model checking*, i.e. stepwise unrolling the transition relation of a system interspersed with model checking runs, facilitates to reuse and propagate knowledge from previous runs (due to symmetries) such as *conflict clause reusing and shifting*. In the quantified case, we are also interested in maintaining and propagating knowledge about *solutions* of previous solver calls. Besides skipping branches leading to a conflict, such a technique would allow to avoid investigation of branches for which the satisfaction probability was (partially) computed previously.
- The underlying iSAT algorithm which is based on interval arithmetic is in general neither able to find a (real) solution nor to prove its absence. In such cases the results are approximative solutions which suffice certain consistency notions but do not guarantee real solutions. Concerning the reliability of the computed satisfaction probabilities, we will extend the SiSAT tool to deal with confidence intervals in order to obtain *safe approximations of satisfaction probabilities*.
- Another issue concerning the reliability of the computed results is to offer *certificates of the satisfaction probabilities*, i.e. proofs that the returned probabilities are correct. Reliable results are of utmost importance in the

verification of safety-critical industrial systems. A certificate that a quantifier-free formula is satisfiable is simply a satisfying assignment to the variables. A proof of unsatisfiability is often more complex, e.g. a clause resolution strategy in the propositional case. In our setting, such a certificate seems to be a mixed version of both.

- To assess the practical significance, we will apply the SiSAT tool on *real-world benchmarks*. Within the AVACS project[1], we are especially interested in benchmarks which deal with the impact of cooperative, distributed traffic management on flow of road traffic. These benchmarks are representative for a large number of hard scheduling and allocation problems and naturally show uncertain behavior.
- A more fundamental challenge is to generalize the scope of the quantifiers to continuous domains involving arbitrary probability distributions. This would increase considerably the expressive power of SSMT.

6   Interpolation

In system's verification, i.e. unbounded model checking, Craig interpolants [25] have gained more and more attention over the last years. In [26], McMillan modified a bounded model checking procedure for Kripke Structures with the help of interpolants s.t. the procedure becomes able to prove safety properties of a given system for runs of *any* length. More recently, McMillan extended his work on unbounded model checking to the quantifier-free theory of linear inequality and uninterpreted function symbols [27], which is used, e.g., in software verification. His approach has been implemented in the software model checker BLAST [28].

As outlined in Section 1, bounded model checking aims at *disproving* a property $P(\boldsymbol{x})$ of a given system $\mathcal{S}$. Thus, a BMC procedure tries to find an unwinding depth k s.t. the corresponding BMC formula Φ_k with $TARGET(\boldsymbol{x}_k) = \neg P(\boldsymbol{x}_k)$ is satisfiable. On the other hand, the goal of *unbounded* model checking is to prove that P holds for all runs of $\mathcal{S}$. I.e., Φ_k with target $\neg P(\boldsymbol{x}_k)$ is unsatisfiable for any $k \in \mathbb{N}$.

Such an unbounded model checking procedure can be obtained by means of *Craig interpolation*. Given two formulae A and B s.t. $A \wedge B$ is unsatisfiable. Then, a formula p is called *Craig interpolant for A and B* iff (1) p contains only variables which occur in both A and B ((A, B)-*common variables*), (2) $A \Rightarrow p$, and (3) $p \Rightarrow \neg B$. A Craig interpolant p is called *strongest* if p implies any other Craig interpolant p', i.e. $p \Rightarrow p'$. Hence, any such interpolant p' overapproximates p.

After showing that $\Phi_0 = INIT(\boldsymbol{x}_0) \wedge TARGET(\boldsymbol{x}_0)$ is unsatisfiable (i.e. initially the target does not hold), McMillan's procedure first solves the BMC formula $\Phi_1 = PREF \wedge SUFF$, where

$$PREF := REACH(\boldsymbol{x}_0) \wedge TRANS(\boldsymbol{x}_0, \boldsymbol{x}_1) \text{ and}$$
$$SUFF := TARGET(\boldsymbol{x}_1),$$

---

[1] www.avacs.org

and initially $REACH(\boldsymbol{x_0}) := INIT(\boldsymbol{x_0})$. If Φ_1 is unsatisfiable then a Craig interpolant $p(\boldsymbol{x_1})$ for the formulae $PREF$ and $SUFF$ is computed.[2] By $PREF \Rightarrow p(\boldsymbol{x_1})$, the interpolant $p(\boldsymbol{x_1})$ is an overapproximation of the states reachable in one system step from $REACH(\boldsymbol{x_0})$. If this overapproximation shifted to the zeroth instantiation of the variables (as described by $p(\boldsymbol{x_0})$) is a subset of the so far reachable states, i.e. $p(\boldsymbol{x_0}) \Rightarrow REACH(\boldsymbol{x_0})$, then further transitions can only lead to states already characterized by $REACH(\boldsymbol{x})$. As a consequence, the target states are unreachable and the verification procedure succeeds. Otherwise, we expand the set of reachable states s.t. it also covers the reachable states given by the shifted interpolant, i.e. $REACH(\boldsymbol{x_0}) := REACH(\boldsymbol{x_0}) \vee p(\boldsymbol{x_0})$. Then, the procedure is iterated until the above termination criterion holds. Due to the overapproximations of the reachable state set, showing the satisfiability of one of the obtained formulae Φ_1 does not imply that the target state is actually reachable. For a more detailed account, confer [26].

Computing Craig interpolants for different theories can be found in the literature [27,29,30]. However, none of these approaches is capable of constructing interpolants for the case of mixed Boolean and non-linear arithmetic constraints including transcendental functions. Therefore, extending the concept of interpolation to this richer domain originating from hybrid systems analysis is an interesting research issue. In the sequel, we identify the most essential challenges:

- Obtaining interpolants in the iSAT case requires construction rules. One way might be to generalize Pudlák's algorithm [31], that delivers interpolants for the propositional case using the proof of unsatisfiability.
- Craig interpolants are not unique and therefore there exist interpolants that are bigger or smaller, stronger or weaker then others, etc. Thus, an open problem is to determine which characteristics of interpolants are favorable especially in the sense of low computational costs.
- As the reachability problem of hybrid systems is in general undecidable, it is worthwhile to identify decidable subclasses for which the interpolation procedure always terminates. One promising starting point is to investigate *robustness* notions for hybrid systems (cf. e.g. [1]), which may guarantee such a termination property.

The following example illustrates that selecting suitable Craig interpolants is a difficult problem. The system $\mathcal{S}$ with $INIT(\boldsymbol{x_0}) = x_0 \geq 1$, $TRANS(\boldsymbol{x_i}, \boldsymbol{x_{i+1}}) = x_{i+1} \geq 0.5x_i$, and $TARGET(\boldsymbol{x_k}) = x_k \leq 0$ describes the evolution of a variable x that is initially greater than 1 and is iteratively divided by 2. A property of $\mathcal{S}$ is that x will never become less than 0. Consider the formula $\Phi_1 = PREF \wedge SUFF = (x_0 >= 1 \wedge x_1 = 0.5x_0) \wedge (x_1 \leq 0)$ which is unsatisfiable. A possible Craig interpolant is $p^1 = x_1 \geq 0.5$. As $p^1 \not\Rightarrow INIT(\boldsymbol{x_0})$ we use p^1 as the new initial state. The resulting formula $(x_0 >= 0.5 \wedge x_1 = 0.5x_0) \wedge (x_1 \leq 0)$ is also unsatisfiable. A possible Craig interpolant is $p^2 = x_1 \geq 0.25$. Since p^2 does not imply p^1, we proceed. If we had computed $p^1 = x_1 > 0$ then p^2 would imply p^1, resulting in a fixed point. Though the example suggests that weaker interpolants are more suitable than stronger ones, this needs not to be true in general.

---

[2] Note that $p(\boldsymbol{x_1})$ may only contain $(PREF, SUFF)$-common variables.

7 Parallelization

Recent trends in hardware design towards multi-core and multiprocessor systems, and computer clusters call for the development of dedicated parallel algorithms in order to exploit the full potential of these architectures. In the domain of propositional SAT solving, parallel algorithms can be traced back to at least 1994, when Böm and Speckenmeyer presented the first parallel implementation of a DPLL procedure for a transputer system consisting of up to 256 processors [32]. During the past decade, more advanced implementations have been developed. Most existing parallel SAT solvers are based on DPLL, they are, however, parallelized in different ways and focus on different hardware environments.

While some of them, such as PaSAT [33], PaMira [34], Satz [35], are designed for distributed memory systems, others, like ySAT [36], MiraXT [37], are tailored to use shared memory workstations. Both shared-memory and distributed-memory workstations have advantages and disadvantages. Shared memory computers have the benefit that all processors can access a shared common address space and guarantee in general low latency and low communication overhead. In distributed systems, on the other hand, each processor has its own local memory. Hence, processors communicate over the network via messages causing slow inter-process communication. Choosing the right memory architecture has thus an important impact on the performance of any parallelized algorithm.

As iSAT builds upon DPLL, adapting different parallelization approaches from purely propositional SAT solving to this richer framework constitutes an important first goal. In [38], *guiding paths* are used to partition the search space of a propositional SAT problem into non-overlapping parts that can be treated in parallel by dynamically allotting them to different processors. The underlying idea is to split the search space at the decision points of the DPLL search tree, i.e. at points where a value for a propositional variable is selected. For this purpose, the guiding path keeps track of possible alternative decisions that can be given to an idle processor. This concept can be adapted to the iSAT context by partitioning the search space at interval-splitting points (cf. Fig. 2).

Furthermore, the exchange of conflict clauses is an essential ingredient of parallel SAT solvers to gain performance. Each conflict clause describes a part of

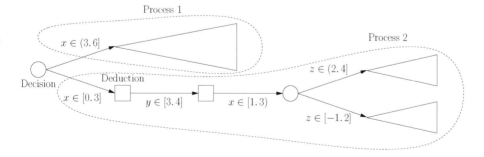

Fig. 2. Search space partitioning at interval splitting points in iSAT (two processors)

the search space which does not contain any solution. Thus, exchanging conflict clauses prevents other processes from examining such conflicting parts that have already been proven unsatisfiable by another process. Another such element is to employ different decision heuristics for each involved processor. In [34], it was shown that selecting the variables according to different decision heuristics accelerated the PaMira solver by 70% on average.

In addition to parallelization techniques from propositional SAT, the iSAT algorithm introduces new options. As the deduction mechanisms in iSAT (e.g. ICP) are in general more expensive than Boolean deductions, parallelizing iSAT's deduction phase could be beneficial. Another observation is that smaller values for the *minimum splitting width* (cf. Section 2) typically cause longer runtimes of iSAT but allow more precise results. Exploiting this, solver instances with greater minimum splitting widths could supply those instances having smaller widths with conflict clauses in order to accelerate their search.

The high computational costs of checking large BMC instances call for the development of parallel BMC techniques. While some approaches apply parallel solvers to the same BMC instance, the authors of [39] introduce a different approach by simultaneously solving different BMC instances. Moreover, they also adapt the concept of sharing and shifting conflict clauses, first proposed by Strichman [40] for sequential BMC, to the parallel setting. Since BMC formulae Φ_k and Φ_m for the same system share common subformulae, it makes sense to exchange conflict clauses between the corresponding solver instances. Shifting conflict clauses is a related technique, that exploits the symmetry between different BMC formulae originating from the same system. As a BMC formula Φ_k consists of k instantiations of the transition relation, conflict clauses can be shifted within the current instantiation depth k. It is an open question whether a similar parallel BMC scheme for non-linear hybrid systems with iSAT as the underlying constraint solver yields performance gains comparable to those encountered for linear hybrid systems using a combined SAT-LP solver [39].

8   Conclusion

In this paper, we sketched a number of challenges emerging from ongoing work on the constraint-based analysis of hybrid systems. While these extensions are currently developed separately from each other, core technologies like ICP or conflict analysis are used commonly. We hope that by keeping these developments closely together, in particular by sharing data structures, synergies become accessible in the long run. We think that among the issues emerging from integration, some are more obvious than others. For example, employing ODE deduction mechanisms as a subordinate solver within SiSAT or the parallelized iSAT seems to be unproblematic. The same holds for the usage of a parallelized solver as a decision engine within the stochastic SMT algorithm or the interpolation approach. On the other hand, even the theory of interpolation within a probabilistic environment is still unclear, as is the generation of interpolants from formulae comprising ODE constraints.

While some details presented in this paper are specific to the iSAT context, we think that the broader issues are of more general interest. For instance, ODE propagation could be used within other SMT approaches [8] as a theory solver, while e.g. decision heuristics and certificate generation may not only be applicable to SSMT but could also be used in stochastic constraint programming [41].

Acknowledgements

The authors would like to thank Erika Ábrahám, Bernd Becker, Christian Herde, Holger Hermanns, and Tobias Schubert for many valuable discussions about the presented topics, and Martin Fränzle for additionally commenting on earlier versions of this paper. Furthermore, the authors are very grateful to the anonymous reviewers for their helpful remarks.

References

1. Fränzle, M.: Analysis of hybrid systems: An ounce of realism can save an infinity of states. In: Flum, J., Rodríguez-Artalejo, M. (eds.) CSL 1999. LNCS, vol. 1683, pp. 126–140. Springer, Heidelberg (1999)
2. Henzinger, T.A., Kopke, P.W., Puri, A., Varaiya, P.: What's decidable about hybrid automata? In: Proc. of the 27th Annual Symposium on Theory of Computing, pp. 373–382. ACM Press, New York (1995)
3. Groote, J.F., Koorn, J.W.C., van Vlijmen, S.F.M.: The Safety Guaranteeing System at Station Hoorn-Kersenboogerd. In: Conference on Computer Assurance, pp. 57–68. National Institute of Standards and Technology (1995)
4. Biere, A., Cimatti, A., Zhu, Y.: Symbolic model checking without BDDs. In: Cleaveland, W.R. (ed.) TACAS 1999. LNCS, vol. 1579, p. 193. Springer, Heidelberg (1999)
5. Audemard, G., Bozzano, M., Cimatti, A., Sebastiani, R.: Verifying industrial hybrid systems with MathSAT. In: Bounded Model Checking (BMC 2004). ENTCS, vol. 119, pp. 17–32 (2005)
6. Fränzle, M., Herde, C.: HySAT: An efficient proof engine for bounded model checking of hybrid systems. Formal Methods in System Design 30(3), 179–198 (2007)
7. Benhamou, F., Granvilliers, L.: Continuous and interval constraints. In: Rossi, F., van Beek, P., Walsh, T. (eds.) Handbook of Constraint Programming. Foundations of Artificial Intelligence, pp. 571–603. Elsevier, Amsterdam (2006)
8. Barrett, C., Sebastiani, R., Seshia, S., Tinelli, C.: Satisfiability Modulo Theories. In: Handbook on Satisfiability, February 2009. Frontiers in Artificial Intelligence and Applications, vol. 185. IO Press (2009), ftp://ftp.cs.uiowa.edu/pub/tinelli/papers/BarSST-09.pdf
9. Bauer, A., Pister, M., Tautschnig, M.: Tool-support for the analysis of hybrid systems and models. In: Design, Automation and Test in Europe. IEEE, Los Alamitos (2007)
10. Fränzle, M., Herde, C., Teige, T., Ratschan, S., Schubert, T.: Efficient Solving of Large Non-linear Arithmetic Constraint Systems with Complex Boolean Structure. JSAT Special Issue on SAT/CP Integration 1, 209–236 (2007)
11. Tseitin, G.: On the complexity of derivations in propositional calculus. Studies in Constructive Mathematics and Mathematical Logics (1968)

12. Davis, M., Putnam, H.: A Computing Procedure for Quantification Theory. Journal of the ACM 7(3), 201–215 (1960)
13. Davis, M., Logemann, G., Loveland, D.: A Machine Program for Theorem Proving. Communications of the ACM 5, 394–397 (1962)
14. Moore, R.E.: Automatic local coordinate transformation to reduce the growth of error bounds in interval computation of solutions of ordinary differential equations. In: Ball, L.B. (ed.) Error in Digital Computation, vol. II, pp. 103–140. Wiley, New York (1965)
15. Lohner, R.: Einschließung der Lösung gewöhnlicher Anfangs- und Randwertaufgaben. PhD thesis, Fakultät für Mathematik der Universität Karlsruhe, Karlsruhe (1988)
16. Stauning, O.: Automatic Validation of Numerical Solutions. PhD thesis, Technical University of Denmark, Lyngby (1997)
17. Berz, M., Makino, K.: Verified Integration of ODEs and Flows Using Differential Algebraic Methods on High-Order Taylor Models. Reliable Computing 4(4), 361–369 (1998)
18. Henzinger, T.A., Horowitz, B., Majumdar, R., Wong-Toi, H.: Beyond HYTECH: Hybrid systems analysis using interval numerical methods. In: Lynch, N.A., Krogh, B.H. (eds.) HSCC 2000. LNCS, vol. 1790, pp. 130–144. Springer, Heidelberg (2000)
19. Hickey, T., Wittenberg, D.: Rigorous modeling of hybrid systems using interval arithmetic constraints. In: Alur, R., Pappas, G.J. (eds.) HSCC 2004. LNCS, vol. 2993, pp. 402–416. Springer, Heidelberg (2004)
20. Eggers, A., Fränzle, M., Herde, C.: SAT modulo ODE: A direct SAT approach to hybrid systems. In: Cha, S(S.), Choi, J.-Y., Kim, M., Lee, I., Viswanathan, M. (eds.) ATVA 2008. LNCS, vol. 5311, pp. 171–185. Springer, Heidelberg (2008)
21. Fränzle, M., Hermanns, H., Teige, T.: Stochastic Satisfiability Modulo Theory: A Novel Technique for the Analysis of Probabilistic Hybrid Systems. In: Egerstedt, M., Mishra, B. (eds.) HSCC 2008. LNCS, vol. 4981, pp. 172–186. Springer, Heidelberg (2008)
22. Papadimitriou, C.H.: Games against nature. J. Comput. Syst. Sci. 31(2), 288–301 (1985)
23. Schmitt, C.: Bounded Model Checking of Probabilistic Hybrid Automata. Master's thesis, Carl von Ossietzky University, Dpt. of Computing Science, Oldenburg, Germany (March 2008)
24. Teige, T., Fränzle, M.: Stochastic Satisfiability modulo Theories for Non-linear Arithmetic. In: Perron, L., Trick, M.A. (eds.) CPAIOR 2008. LNCS, vol. 5015, pp. 248–262. Springer, Heidelberg (2008)
25. Craig, W.: Linear reasoning: A new form of the Herbrand-Gentzen theorem. Journal of Symbolic Logic 22(3), 250–268 (1957)
26. McMillan, K.L.: Interpolation and SAT-based model checking. In: Hunt Jr., W.A., Somenzi, F. (eds.) CAV 2003. LNCS, vol. 2725, pp. 1–13. Springer, Heidelberg (2003)
27. McMillan, K.L.: An interpolating theorem prover. Theor. Comput. Sci. 345(1), 101–121 (2005)
28. Beyer, D., Henzinger, T.A., Jhala, R., Majumdar, R.: The software model checker Blast: Applications to software engineering. International Journal on Software Tools for Technology Transfer (STTT) 9(5-6), 505–525 (2007) (invited to special issue of selected papers from FASE 2004/2005)
29. Rybalchenko, A., Sofronie-Stokkermans, V.: Constraint solving for interpolation. In: Cook, B., Podelski, A. (eds.) VMCAI 2007. LNCS, vol. 4349, pp. 346–362. Springer, Heidelberg (2007)

30. Cimatti, A., Griggio, A., Sebastiani, R.: Efficient interpolant generation in satisfiability modulo theories. In: Ramakrishnan, C.R., Rehof, J. (eds.) TACAS 2008. LNCS, vol. 4963, pp. 397–412. Springer, Heidelberg (2008)
31. Pudlák, P.: Lower Bounds for Resolution and Cutting Plane Proofs and Monotone Computations. Journal of Symbolic Logic 62(3), 981–998 (1997)
32. Böhm, M., Speckenmeyer, E.: A fast parallel SAT-solver - efficient workload balancing. Annals of Mathematics and Artificial Intelligence 17(3-4), 381–400 (1996)
33. Sinz, C., Blochinger, W., Küchlin, W.: PaSAT - parallel SAT-checking with lemma exchange: Implementation and applications. In: Kautz, H., Selman, B. (eds.) LICS 2001 Workshop on Theory and Applications of Satisfiability Testing (SAT 2001), June 2001. Electronic Notes in Discrete Mathematics, vol. 9. Elsevier Science Publishers, Boston (2001)
34. Schubert, T., Lewis, M., Becker, B.: PaMira – A Parallel SAT Solver with Knowledge Sharing. In: 6th International Workshop on Microprocessor Test and Verification (MTV 2005), pp. 29–36. IEEE Computer Society, Los Alamitos (2005)
35. Jurkowiak, B., Li, C.M., Utard, G.: Parallelizing SATZ Using Dynamic Workload Balancing. In: Proceedings of the Workshop on Theory and Applications of Satisfiability Testing (SAT 2001), June 2001, vol. 9. Elsevier Science Publishers, Amsterdam (2001)
36. Feldman, Y., Dershowitz, N., Hanna, Z.: Parallel multithreaded satisfiability solver: Design and implementation. Electronic Notes in Theoretical Computer Science 128(3), 75–90 (2005)
37. Lewis, M.D.T., Schubert, T., Becker, B.: Multithreaded SAT solving. In: Proceedings of the 12th Asia and South Pacific Design Automation Conference, pp. 926–931. IEEE Computer Society, Los Alamitos (2007)
38. Zhang, H., Bonacina, M.P., Hsiang, J.: PSATO: A distributed propositional prover and its application to quasigroup problems. Journal of Symbolic Computation 21(4), 543–560 (1996)
39. Ábrahám, E., Schubert, T., Becker, B., Fränzle, M., Herde, C.: Parallel SAT solving in bounded model checking. In: Brim, L., Haverkort, B.R., Leucker, M., van de Pol, J. (eds.) FMICS 2006 and PDMC 2006. LNCS, vol. 4346, pp. 301–315. Springer, Heidelberg (2007)
40. Strichman, O.: Accelerating bounded model checking of safety properties. Formal Methods in System Design 24(1), 5–24 (2004)
41. Walsh, T.: Stochastic constraint programming. In: Proc. of the 15th European Conference on Artificial Intelligence (ECAI 2002). IOS Press, Amsterdam (2002)

From Rules to Constraint Programs with the Rules2CP Modelling Language

François Fages and Julien Martin

Projet Contraintes, INRIA Rocquencourt,
BP105, 78153 Le Chesnay Cedex, France
http://contraintes.inria.fr

Abstract. In this paper, we present a rule-based modelling language for constraint programming, called Rules2CP. Unlike other modelling languages, Rules2CP adopts a single knowledge representation paradigm based on rules without recursion, and a restricted set of data structures based on records and enumerated lists given with iterators. We show that this is sufficient to model constraint satisfaction problems, together with search strategies where search trees are expressed by logical formulae, and heuristic choice criteria are defined by preference orderings on variables and formulae. We describe the compilation of Rules2CP statements to constraint programs over finite domains, by a term rewriting system and partial evaluation. We prove the confluence of these transformations and provide a complexity bound on the size of the generated programs. The expressiveness of Rules2CP is illustrated first with simple examples, and then with a complete library for packing problems, called PKML, which, in addition to pure bin packing and bin design problems, can deal with common sense rules about weights, stability, as well as specific packing business rules. The performances of both the compiler and the generated code are evaluated on Korf's benchmarks of optimal rectangle packing problems.

1 Introduction

From a programming language standpoint, one striking feature of constraint programming is its declarativity for stating combinatorial problems, describing only the "what" and not the "how", and yet its efficiency for solving large size problem instances in many practical cases. From an application expert standpoint however, constraint programming is not as declarative as one would wish, and constraint programming systems are in fact very difficult to use by non-programmers outside the range of already treated examples. This well recognized difficulty has been presented as a main challenge for the constraint programming community, and has motivated the search for more declarative front-end problem modelling languages, such as for instance OPL [1,2], Zinc [3,4] and Essence [5].

In industry, the business rules approach to knowledge representation has a wide audience because of the declarativity and granularity of rules which can be introduced, checked, and modified one by one, and independently of any

A. Oddi, F. Fages, and F. Rossi (Eds.): CSCLP 2008, LNAI 5655, pp. 66–83, 2009.

particular procedural interpretation by a rule engine [6]. This provides an attractive knowledge representation scheme for quickly evolving requirements, and for maintaining systems with up to date information. In this article, we show that such a rule-based knowledge representation paradigm can be developed as a front-end modelling language for constraint programming. We present a general purpose rule-based modelling language for constraint programming, called Rules2CP. Unlike multi-headed condition-action rules, also called production rules, Rules2CP rules are restricted to *logical rules*, with one head and no imperative actions, and where bounded quantifiers are used to represent complex conditions. Such rules comply to the principle of independence from a procedural interpretation by a rule engine [6], which is concretely demonstrated in Rules2CP by their compilation to constraint programs using a completely different representation.

Unlike the other modelling languages proposed for constraint programming, Rules2CP adopts a restricted set of data structures based on records and enumerated lists, given with iterators. We show that this is sufficient to express constraint satisfaction problems, together with search strategies where the search tree is expressed by logical formulae, and complex heuristic choice criteria are defined as preference orderings on variables and formulae.

The next section presents the Rules2CP language and shows how search strategies and heuristics can be specified in a declarative manner. Sec. 2 describes the compilation of Rules2CP models into constraint programs over finite domains with reified constraints, by term rewriting and partial evaluation. We prove the confluence of these transformations which shows that the generated constraint program does not depend on the order of application of the rewritings, and provide a complexity bound on the size of the generated program.

Sec. 4 illustrates the expressive power of this approach with a particular Rules2CP library, called the Packing Knowledge Modelling Library (PKML), developed in the EU project Net-WMS[1] for dealing with real-size non-pure bin packing problems of the automotive industry. The performances of both the compiler and the generated code are evaluated in this section on Korf's benchmarks of optimal rectangle packing [7].

Finally, Sec. 5 compares Rules2CP with related work on OPL, Zinc and Essence modelling languages, business rules, constraint logic programming and term rewriting systems. We conclude on the simplicity and efficiency of Rules2CP and on some of its current limitations.

2   The Rules2CP Language

2.1   Introductory Examples

Rules2CP is an untyped language for modelling constraint satisfaction problems over finite domains using rules and declarations with records and enumerated lists as data structures. Let us first look at some simple examples.

---

[1] http://net-wms.ercim.org

Example 1. The classical N-queens problem, i.e. placing N queens on a chess-board of size $N \times N$ such that the queens are not on the same raw, column or diagonal, can be modelled in Rules2CP with two declarations (q and board), for creating a list of records representing the positions of the queens on the chess board, one rule safe for defining when the queens do not attack each other (using the global constraint all_different below), another rule solve for defining the constraints and the search strategy, and one goal for solving a problem of a given size:

```
q(I) = {row = _, column = I}.
board(N) = map(I, [1..N], q(I)).
safe(B) --> all_different(B) and
          forall(Q, B, forall(R, B,
          let(I, column(Q), let(J, column(R),
            I<J implies row(Q)#J-I+row(R) and row(Q)#I-J+row(R)))))).
solve(N) --> let(B, board(N), domain(B, 1, N) and safe(B) and
                  dynamic(variable_ordering([least(domain_size(row(^)))])
                      and labeling(B))).
? solve(4).
```

The search is specified in the solve rule by the labeling predicate for enumerating the variables contained in B with a *dynamic* variable ordering heuristics by least domain size (first-fail heuristics).

Example 2. A disjunctive scheduling problem, such as the classical bridge problem [1], consists in finding the earliest start dates for a set of tasks given with their durations, under constraints of precedence and mutual exclusion between tasks. Such problems can be modelled in Rules2CP with records for tasks, and rules for precedence and disjunctive constraints, as follows:

```
t1 = {start=_, duration=2}. t2 = {start=_, duration=5}.
t3 = {start=_, duration=4}. t4 = {start=_, duration=3}.
t5 = {start=_, duration=1}.
prec(T1, T2) --> start(T1) + duration(T1) =< start(T2).
disj(T1, T2) --> prec(T1,T2) or prec(T2,T1).
precedences --> prec(t1,t2) and prec(t2,t5) and prec(t1,t3) and prec(t3,t5)
disjunctives --> disj(t2,t3) and disj(t2,t4) and disj(t3,t4).
? domain([t1,t2,t3,t4,t5], 0, 20) and precedences and
  conjunct_ordering([greatest(duration(A)+duration(B) if ^ is disj(A,B))])
  and minimize(disjunctives, start(t5)).
```

The goal posts the domain and precedence constraints, specifies a heuristic criterion for ordering the disjunctive constraints by decreasing durations of tasks, and defines the search strategy by a logical formula (disjunctives) composed of a conjunction of disjunctive constraints, and a minimization criterion (the starting date of task t6). It is worth noting that this model does not use variable labeling. In a computed optimal solution, the non-critical tasks will have a flexible starting date.

The ordering criterion is about the duration attributes of the tasks involved in the disj rules, and does not actually depend on the variables. This strategy corresponds to the ordering used implicitly in the classical bridge problem benchmark. By adding a criterion for selecting the disjunctive with highest difference of durations in case of equality, as follows

```
conjunct_ordering([greatest(duration(A)+duration(B) if ^ is disj(A,B)),
                greatest(abs(duration(A)-duration(B)) if ^ is disj(A,B))]).
```

the performances are slightly improved in the bridge problem.

2.2 Syntax

Let an *ident* be a word beginning with a lower case letter or any word between quotes, a *name* be an identifier possibly prefixed by other identifiers for module and package names, and a *variable* be a word beginning with either an upper case letter or the underscore character. The syntax of Rules2CP statements is given by the following grammar:

statement ::=	import *name*. | *head* = *expr*. | *head* --> *fol*. | ? *fol*.
name ::=	*ident* | *name*:*ident*
head ::=	*ident* | *ident(var,...,var)*
fol ::=	*varbool* | *name* | *name(expr,...,expr)* | *expr relop expr*
	| *fol logop fol* | not *fol* | forall*(var,expr,fol)* | exists*(var,expr,fol)*
	| foldl*(var,expr,logop,expr,expr)* | foldr*(var,expr,logop,expr,expr)*
	| let*(var,expr,fol)* | search*(fol)* | dynamic*(fol)*
expr ::=	*varint* | *fol* | *string* | [*enum*] | {*ident* = *expr*,...,*ident*= *expr*}
	| *name* | *name(expr,...,expr)* | *expr op expr* |
	| foldl*(var,expr,op,expr,expr)* | foldr*(var,expr,op,expr,expr)*
	| map*(var,expr,expr)*
enum ::=	*enum* , *enum* | *expr* | *expr* .. *expr*
varint ::=	*var* | *integer*
varbool ::=	*var* | 0 | 1
op ::=	+ | − | * | / | min | max | log | exp
relop ::=	< | =< | = | # | >= | >
logop ::=	and | or | implies | equiv | xor

A statement is either a module import, a declaration, a rule or a goal. In order to avoid name clashes in declaration and rule heads, the language includes a simple module system that prefixes names with module and package names, similarly to [8]. A head is formed with an *ident* with distinct variables as arguments. Recursive definitions, as well as multiple definitions of a same head symbol, are forbidden in declarations and rules, and each name must be defined before its use. Apart from this, the order of the statements in a Rules2CP file is not relevant.

The set $V(E)$ of *free variables* in an expression E is the set of variables occurring in E and not bound by a forall, exists, let, foldl, foldr or map operator. In a rule, $L\mathtt{-->}R$, we assume $V(R) \subseteq V(L)$, whereas in a declaration, $H\mathtt{=}E$, the introduced variables, in $V(E) \setminus V(H)$, represent unknown variables of the problem.

The only data structures are integers, strings, enumerated lists and records. Lists are formed without a binary list constructor, by enumerating all their elements, or intervals of values in the case of integers. For instance $\mathtt{[1,3..6,8]}$ represents the list $\mathtt{[1,3,4,5,6,8]}$. Such lists are used to represent the domains of variables in *(var in list)* formula, and in the answers returned to Rules2CP goals. The following expressions: $\mathtt{length}$(*list*), $\mathtt{nth}$(*integer,list*), $\mathtt{pos}$(*element,list*) and *attribute*(*record*) are predefined for accessing the components of lists and records. Furthermore, records have a default integer attribute $\mathtt{uid}$ which provides them with a unique identifier.

The predefined function $\mathtt{variables}$(*expr*) returns the list of variables contained in an expression. The predefined predicate X $\mathtt{in}$ *list* constrains the variable X to take integer values in a list of integer values. $\mathtt{domain}$(*expr,min,max*)) is predefined to set the domain of all variables occurring in *expr*.

A *fol* formula can be considered as a 0/1 integer expression. This usual coercion between booleans and integers, called *reification*, provides a great expressivity [9]. The (left and right) fold operators cannot be defined in first-order logic and are Rule2CP builtins. These operators iterate the application of a binary operator on a list of arguments. For instance, the product of the elements in a list is defined by $\mathtt{product(L)=foldr(X,L,*,1,X)}$. Furthermore, a *fol* formula can be evaluated dynamically instead of statically by prefixing the formula with the predicate $\mathtt{dynamic}$.

2.3   Search Predicates

Describing the search strategy in a modelling language is a challenging task as search is usually considered as inherently procedural, and thus contradictory to declarative modelling. This is however not our point of view in Rules2CP. Our approach to this issue is to specify the *decision variables* and the *branching formulas* of the problem in a declarative manner, and then heuristics as *preference orderings* on variables and formulae.

In Rules2CP, the labeling of decision variables can be specified with the predefined predicate $\mathtt{labeling}$(*expr*) for enumerating the possible values of all the variables *contained* in an expression, that is occurring as attributes of a record, or recursively in a record referenced by attributes, in a list, or in a first-order formula (see Example 1). The *branching formulas* are declared similarly with the predicate $\mathtt{search}$(*fol*) for specifying a search procedure by branching on all disjunctions and existential quantifications occurring in a first-order formula (see Example 2). Note that without the search predicate, the formula in argument would be treated as a constraint by reification. A similar approach to specifying search has been proposed for SAT in [10]. Here however, the only normalization is the elimination of negations in the formula by descending them to the constraints. The structure of the formula is kept as an *and-or* search tree where the disjunctions constitute the choice points.

The predefined *optimisation predicates*, `minimize(`*fol,expr*`)` for searching a fol and minimizing an expression, and `maximize(`*fol,expr*`)`, can be imbricated. This makes it possible to express multicriteria optimisation problems, and the search for Pareto optimal solutions according to the lexicographic ordering of the criteria as read from left to right.

2.4 Heuristics as Ordering Criteria

Adding the capability to express heuristic knowledge is mandatory for efficiency. This is done in Rules2CP with predefined predicates for specifying both static and dynamic choice criteria on variables and values for `labeling`, and on conjunctive and disjunctive formulae for `search`. Dynamic criteria for ordering variables and values in `labeling` are standard in constraint programming systems, see for instance [11,12]. In Rules2CP, they are defined more generally using the expressive power of the language for specifying various criteria depending on static or dynamic expression values.

The `variable_ordering` predicates take a list of criteria for ordering the variables in subsequent `labeling` predicate. The variables are sorted according to the first criterion when it applies, then the second, etc. The variables for which no criterion applies are considered at the end for labeling in the syntactic order. The criteria have the following forms: `greatest(`*expr*`)`, `least(`*expr*`)`, `any(`*expr*`)` or `is(`*expr*`)`. The expression *expr* in a criterion contains the symbol `^` for denoting, for a given variable, the left-hand side of the Rules2CP declaration that introduced the variable. If the expression cannot be evaluated on a variable, the criterion is ignored. An `any` form selects a variable for which the expression applies, independently of its value. An `is` form selects a variable if it is equal to the result of the expression. For instance, in a 3-dimensional bin packing problem, the predicate `variable_ordering([greatest(volume(^)), least(uid(^))])` specifies a lexicographic static ordering of the variables by decreasing volume of the object in which they have been declared, and by increasing *uid* attribute of the object (for grouping the variables belonging to a same object).

The `value_ordering` predicate takes similarly a list of criteria of the forms: `up`, `up(`*expr*`)`, for enumerating values in ascending order for the variables matching the expression, or `down`, `step` for binary choices, `enum` for multiple choices, `bisect` for dichotomy choices. A criterion applies to a variable if it matches the expression. For instance, in a bin packing problem with x, y, z coordinates, the predicate `value_ordering([up(z(^)), bisect(x(^)), bisect(y(^))])` specifies the enumeration in ascending order for the z coordinates, and by dichotomy for the x and y coordinates. The capabilities of dissociating the specifications of the variable and value heuristics, and of using static criteria about the objects in which the variables appear, are very powerful. It is worth noticing that this expressive power for the heuristics creates difficulties however for their compilation to the constraint programming systems that mix variable and value choice strategies in a single option list, and for which one cannot express different value choice heuristics for the different variables in a same labeling predicate [12]. In these cases, the compiler generates a labeling program.

In search trees defined by logical formulae, the criteria for `conjunct_ordering` and `disjunct_ordering` heuristics are defined similarly by pattern matching on the rule heads that introduce conjunctive and disjunctive formulae under the `search` predicate. This is illustrated in Example 2 with conditional expressions of the form `if ^ is` ϕ; where ^ denotes the conjunct or disjunct candidate for matching ϕ, and ϕ denotes either a rule head or directly a formula. The conjuncts or disjuncts for which no criterion applies are considered last.

3   Compilation to Constraint Programs over Finite Domains with Reified Constraints

Rules2CP models can be compiled to constraint satisfaction problems over finite domains with reified constraints by interpreting Rules2CP statements using a term rewriting system, i.e. with a rewriting process that rewrites subterms inside terms according to general term rewriting rules. Let the *size* of an expression or formula be the number of nodes in its tree representation, and let us denote by $\rightarrow$ the term rewriting rules of the compilation process. These rules are composed of generic rewrite rules and code generation rules.

3.1   Generic Rewrite Rules

The following rewriting rules are associated to Rules2CP declarations and rules:

$L \rightarrow R$ for every rule of the form L `-->` R (where $V(R) \subseteq V(L)$)

$L\sigma \rightarrow R\sigma\theta$ for every declaration of the form $L = R$ and every ground substitution σ of the variables in $V(L)$, where θ is a renaming substitution that gives unique names indexed by $L\sigma$ to the variables in $V(R) \setminus V(L)$.

In a Rules2CP rule, all the free variables of the right-hand side have to appear in the left-hand side. In a declaration, there can be free variables introduced in the right hand side and their scope is global. Hence these variables are given unique names (with substitution θ) which will be the same at each invocation of the declaration. These names are indexed by the left-hand side of the declaration statement which has to be ground in that case (substitution σ). For example, the row variables in the records declared by `q(N)` in Example 1 are given a unique name indexed by the instance of the head $q(i)$. These conventions provide a basic book-keeping mechanism for retrieving the Rules2CP variables introduced in declarations from their variable names. This is necessary to implement the heuristic criteria, as well as for debugging and user-interaction purposes [13].

The ground arithmetic expressions are rewritten with the rule

$expr \rightarrow v$ if $expr$ is a ground expression and v is its value,

This rule provides a *partial evaluation* mechanism for simplifying the arithmetic expressions as well as the boolean conditions. This is crucial to limiting the size of the generated program and eliminating at compile time the potential overhead due to the data structures used in Rules2CP.

The accessors to data structures are rewritten with the following rule schemas that impose that the lists in arguments are expanded first:

$[i \ .. \ j] \ \rightarrow \ [i, \ i+1, \ldots, j]$ if i and j are integers and $i \leq j$
$\texttt{length}([e_1, \ldots, e_N]) \ \rightarrow \ N$
$\texttt{nth}(i, [e_1, \ldots, e_N]) \ \rightarrow \ e_i$
$\texttt{pos}(e, [e_1, \ldots, e_N]) \ \rightarrow \ i$ where e_i is the *first* occurrence of e in the list after
rewriting,
attribute(R) $\rightarrow \ V$ if R is a record with value V for *attribute*.

The quantifiers, `foldr`, `foldl`, `map` and `let` operators are binding operators
which use a dummy variable to denote place holders in an expression. They are
rewritten under the condition that their first argument is a *variable* and their
second argument is an *expanded list*:

$\texttt{foldr}(X, [e_1, \cdots, e_N], op, e, \phi) \ \rightarrow \ \phi[X/e_1] \ op \ (\ldots \ op \ \phi[X/e_N]) \ (e \ \text{if} \ N = 0)$
$\texttt{forall}(X, [e_1, \cdots, e_N], \phi) \ \rightarrow \ \phi[X/e_1] \ \texttt{and} \ \ldots \ \texttt{and} \ \phi[X/e_N] \ (1 \ \text{if} \ N = 0)$
$\texttt{exists}(X, [e_1, \cdots, e_N], \phi) \ \rightarrow \ \phi[X/e_1] \ \texttt{or} \ \ldots \ \texttt{or} \ \phi[X/e_N] \ (0 \ \text{if} \ N = 0)$
$\texttt{map}(X, [e_1, \cdots, e_N], \phi) \ \rightarrow \ [\phi[X/e_1], \ \ldots, \ \phi[X/e_N]]$
$\texttt{let}(X, e, \phi) \ \rightarrow \ \phi[X/e]$

where $\phi[X/e]$ denotes the formula ϕ where each free occurrence of variable X in
ϕ is replaced by expression e (after the usual renaming of the variables in ϕ in
order to avoid name clashes with the free variables in e).

Negations are eliminated by descending them to the variables and comparison
operators, with the obvious duality rules for the logical connectives, such as for
instance, replacing the negation of `and` (resp. `equiv`) into `or` (resp. `xor`) etc. It is
worth noting that these transformations do not increase the size of the formula.

3.2 Code Generation Rules

After the application of the previous generic rewrite rules, the actual transformation of a Rules2CP model to a constraint program of some target language,
is specified with *code generation rules*. Such rules are needed for the terms that
are not defined by Rules2CP statements, e.g. builtin constraints, as well as for
the arithmetic and logical expressions that are not expanded with the generic
rewrite rules described in the previous section. The free variables in declarations
are translated into finite domain variables of the target language, with the basic
book-keeping mechanism provided by the naming conventions.

The examples of code generation rules given in this section concern the compilation of Rules2CP to SICStus-Prolog [12]. Basic constraints are thus rewritten
with term rewriting rules such as the following ones, where backquotes in strings
indicate subexpressions to rewrite:

$A > B \rightarrow$ `"'A #> 'B"`
A and $B \rightarrow$ `"'A #/\ 'B"`
$\texttt{lexicographic}(L) \rightarrow$ `"lex_chain('L)"`
$\texttt{domain}(E, M, N) \rightarrow$ `"domain(L, M, N)"` if M and N are integers and where L is
the list of variables remaining in E after rewriting
$\texttt{minimize}(F, C) \rightarrow$ `"minimize((search('F),labeling([up],'L)),'C)"` where
L is the list of variables occurring in the cost expression C.

Obviously, such code generation rules generate programs of linear size. In addition to this static expansion of Rules2CP goals in a constraint program goal, clauses are also generated for rules and declarations in order to interpret the expressions under `dynamic` with the Rules2CP interpreter, which is not be described for lack of space.

Example 3. The compilation of the N-queens problem in Example 1 generates the following SICStus Prolog program :

```
:- use_module(library(clpfd)).
:- use_module(r2cp).
...
solve([Q_1,Q_2,Q_3,Q_4]) :-
  rcp_var(from(q(1),0,1), Q_1), rcp_var(from(q(2),0,1), Q_2),...
  domain([Q_1,Q_2,Q_3,Q_4], 1, 4),
  all_different([Q_1,Q_2,Q_3,Q_4]),
  Q_1#\=1+Q_2, Q_1#\= -1+Q_2, Q_1#\=2+Q_3, Q_1#\= -2+Q_3, Q_1#\=3+Q_4,
  Q_1#\= -3+Q_4, Q_2#\=1+Q_3, Q_2#\= -1+Q_3, Q_2#\=2+Q_4,
  Q_2#\= -2+Q_4, Q_3#\=1+Q_4, Q_3#\= -1+Q_4,
  rcp_variable_ordering([least(var_order_criterion(1,[]))]),
  rcp_labeling([Q_1,Q_2,Q_3,Q_4]).
```

Note that the inequality constraints are properly posted on ordered pairs of queens, and that the other pairs of queens generated by the universal quantifiers have been eliminated at compile time by partial evaluation. As the search heuristics is dynamic, the Rules2CP interpreter is included in the generated program to interpret the dynamic variable ordering heuristics using the labeling predicate of the Rules2CP interpreter. In this case, the program is equivalent to SICStus Prolog `labeling` with the first-fail option but the method is general.

Example 4. The disjunctive scheduling problem in Example 2 is compiled in a constraint program which does not use the Rules2CP interpreter:

```
solve([T1,T2,T4,T3,T5]) :-
  domain([T1,T2,T3,T4,T5], 0, 20),
  T1+2#=<T2, T2+5#=<T5, T1+2#=<T3, T3+4#=<T5, T1+2#=<T3, T3+4#=<T5,
  minimize(((((T2+5#=<T3;T3+4#=<T2),(T2+5#=<T4;T4+3#=<T2),
            (T3+4#=<T4;T4+3#=<T3)),labeling([up],[T5])), T5).
```

In the `minimize` predicate, the disjunctive formulae in the *and-or* search tree have been reordered according to the heuristics by decreasing sums of the task durations. The labeling of the variables contained in the cost function is added by the compiler.

3.3 Confluence, Termination and Complexity

By having forbidden multiple definitions, and restricted the heads to contain only distinct variables as arguments, one can show :

Proposition 1. *For any* Rules2CP *model, the compilation term rewriting system* → *is confluent.*

This means that the rewriting rules can be applied in any order, and generate the same constraint program on a given input model. The proof in [14] shows that the term rewriting system $\to$ is orthogonal, i.e. left-linear and non-overlapping, which entails confluence [15] without termination assumption. By forbidding recursion however, termination clearly holds:

Definition 1. *Given a Rule2CP model M, let the definition rank $\rho(s)$ of a symbol s be defined inductively by:*

$\rho(s) = 0$ *if s is not the head symbol of a declaration or rule in M,*
$\rho(s) = n + 1$ *if s is the head symbol of a declaration or rule in M, and n is the greatest definition rank of the symbols in the right hand side of its declaration or rule.*

The definition rank of M is the maximum definition rank of the symbols in M.

Proposition 2. *For any* Rules2CP *model, the term rewriting system $\to$ is Noetherian.*

Furthermore, a complexity bound on the size of the generated program can be obtained.

Definition 2. *Given a Rule2CP model M, let the fold rank $\alpha(s)$ of a symbol s be defined inductively by:*

$\alpha(s) = 0$ *if s is not the head symbol of a declaration or rule in M,*
$\alpha(s) = max\{n + \alpha(s') \mid L = R \in M$, s *is the head symbol of L and R contains a nesting of n fold operators or quantifiers on an expression containing symbol $s'\}$.*

The fold rank of M is the maximum fold rank of the symbols in M.

Proposition 3. *For any* Rules2CP *model M, the size of the generated program is in $O(l^a * b^r)$, where l is the maximum length of the lists in M (or at least 1), a is the fold rank of M, b is the maximum size of the declaration and rule bodies in M, and r is the definition rank of M.*

Proof. The proof is by induction on a. In the base case, $a = 0$, there is no fold operator in M, and the size of the generated program is linearly bounded by r duplications of rule bodies, i.e. is in $O(b^r)$. In the induction case, $a > 0$, let us first consider the size of the program generated without rewriting the outermost occurrences of fold and quantifier operators. By induction, this size is in $O(l^{a-1} * b^r)$. Now, this generated program can be duplicated at most l times by the outermost fold operators, hence the total size is in $O(l^a * b^r)$ under this strategy. Since by confluence Prop. 1, the generated program is independent of the strategy, the size of the generated program is thus in $O(l^a * b^r)$ under any strategy.

In the N-queens problem of Example 1, since the fold rank is 2, the proposition thus tells us that the size of the generated program for a board of size l is indeed in $O(l^2)$.

4   The Packing Knowledge Modelling Library PKML

In this section, we illustrate the expressive power of Rules2CP with the definition of a Packing Knowledge Modelling Library (PKML) developed in the Net-WMS project for dealing with real size non-pure bin packing problems in logistics and automotive industry. A large subset of PKML rules restricted to linear constraints has been shown in [16] to be compilable with indexical constraints in the geometrical kernel of the global constraint `geost` [17] for higher-dimensional placement problems. Here we define PKML as a library of Rules2CP declarations and rules.

4.1   Shapes and Objects

PKML refers to shapes in $\mathbb{Z}^K$. A *point* in this space is represented by the list of its K integer coordinates `[i1,...,iK]`. These coordinates may be variables or fixed integer values.

In PKML, a *shape* is a *rigid assembly of boxes*. A *box* is an orthotope in $\mathbb{Z}^K$, and is represented in PKML by a record containing one *size* attribute giving the list of the lengths of the box in each dimension. A shape is represented by a record containing one attribute *boxes* for the list of boxes composing the shape, and one attribute *positions* for the list of their positions in the assembly (i.e. a list of lists of coordinates). For instance:

```
point1 = [x1,...,xK].
box1 = {size = [l1,...,lk]}.
shape1 = { boxes=[b1,...,bM], positions=[p1,...,pM]}
object1 = {shapes=[s1,...,sN], shape=_, origin=[x1,...,xK]}
```

A PKML *object*, such as a bin or an item, is a record containing one attribute shapes for the list of its *alternative shapes*, one *origin* point, and some optional attributes such as weight, virtual reality representations or others. The alternative shapes of an object may be the discrete rotations of a basic shape, or different object shapes in a configuration problem. We do not distinguish between items and bins features, since bins at one level can become items at another level, like for instance in a multilevel bin packing problem for packing items into cartons, cartons in pallets, and pallets into trucks.

The following declarations define respectively the volume of a box, a shape composed of a single box, the size of a shape (i.e. assembly of boxes) in a given dimension, and the volume of a shape in given dimensions (assuming no overlap in the assembly):

```
volume_box(B) = product(size(B)).
box(L) = { boxes = [ {size = L} ], positions = [ map(_,L,0) ] }.
size(S, D) = foldr(I, [1..length(boxes(S))], max, 0,
                    nth(D,nth(I,positions(S))) +
                    nth(D,size(nth(I,boxes(S))))).
volume_assembly(S, Dims) = foldr(B, boxes(S), +, 0, volume_box(B)).
```

It is worth noting that if the sizes of the boxes composing the shapes are known, the size and volume expressions evaluate into fixed integer values, whereas if the sizes are unknown, the expressions evaluate to terms containing variables. These terms are used in PKML to define with reified constraints the end in one dimension and the volume of an object with alternative shapes, as follows:

```
origin(O, D) = nth(D, origin(O)).
end(O, D) = origin(O, D) + foldr(S, shapes(O), +, 0,
                    (shape(O)=pos(S,shapes(O)))*size(S, D)).
volume(O, Dims) = foldr(S, shapes(O), +, 0,
                    (shape(O)=pos(S,shapes(O)))*volume_assembly(S,Dims)).
```

4.2   Placement Relations

PKML uses Allen's interval relations [18] in one dimension, and the topological relations of the Region Connection Calculus [19] in higher-dimensions, to express placement constraints. These relations are predefined in libraries [14]. They are used in PKML to define packing rules for pure bin packing and pure bin design problems, symmetry breaking strategies, as well as specific packing business rules for non pure problems taking into account other common sense rules and industrial requirements and expertise.

The part of the PKML library dealing with pure *bin packing problems* is defined as follows:

```
non_overlapping(Items, Dims) -->
   forall(O1, Items, forall(O2, Items,
       uid(O1) < uid(O2) implies not overlap(O1, O2, Dims))).
containmentAE(Items, Bins, Dims) -->
   forall(I, Items, exists(B, Bins, contains_touch_rcc(B,I,Dims))).
bin_packing(Items, Bins, Dims) -->
   containmentAE(Items, Bins, Dims) and non_overlapping(Items, Dims) and
   labeling(Items).
```

The rules define respectively the non-overlapping of a list of items in a list of dimensions, the containment of all items in bins, and pure bin packing problems. Pure *bin design problems* are defined similarly with a declaration for the volume of a bin, and a containment rule in some bin of all items:

```
containmentEA(Items, Bins, Dims) -->
   exists(B, Bins, forall(I, Items, contains_touch_rcc(B,I,Dims))).

bin_design(Bin, Items, Dims) -->
   containmentEA(Items, [Bin], Dims) and
   minimize(labeling(Items), volume(Bin)).
```

Example 5. On the following simple shape pure bin packing problem

```
import(lib:pkml:pkml).
s1 = box([5,4,4]). s2 = box([4,4,2]). s3 = box([5,4,2]).
o1=object(s1,[0,0,0]). o2=object(s2,[_,_,_]). o3=object(s3,[_,_,_]).
```

```
dimensions = [1,2,3]. bins = [o1]. items = [o2, o3].
? variable_ordering([greatest(volume(^, dimensions)), is(z(^))]) and
  bin_packing(items, bins, dimensions).
```

the compiler generates the following SICStus-Prolog goal where the coordinate variables are statically ordered for labeling:

```
solve([O2,O2_2,O2_3,O3,O3_2,O3_3]) :-
  0#=<O2, O2+4#=<5, 0#=<O2_2, O2_2+4#=<4, 0#=<O2_3, O2_3+2#=<4, 0#=<O3,
  O3+5#=<5, 0#=<O3_2, O3_2+4#=<4, 0#=<O3_3, O3_3+2#=<4,
  O2+4#=<O3#\/O3+5#=<O2#\/(O2_2+4#=<O3_2#\/O3_2+4#=<O2_2
          #\/(O2_3+2#=<O3_3#\/O3_3+2#=<O2_3)),
  labeling([], [O3_3,O3,O3_2,O2_3,O2,O2_2]).
```

4.3   Packing Rules

Packing business rules are defined in Rules2CP to take into account further common sense or industrial requirements that are beyond the scope of pure bin packing problems [20]. For instance, the following rules about weights:

```
gravity(Items) -->
   forall(O1, Items, origin(O1, 3) = 0 or
      exists(O2, Items, uid(O1) # uid(O2) and on_top(O1, O2))).
weight_stacking(Items) -->
   forall(O1,  Items, forall(O2, Items,
      (uid(O1) # uid(O2) and on_top(O1, O2)) implies lighter(O1,O2))).
weight_balancing(Items, Bin, D, Ratio) -->
 let(L, sum( map(Il, Items, weight(Il)*(end(Il,D) =< (end(Bin,D)/2)))),
   let(R, sum( map(Ir, Items, weight(Ir)*(end(Ir,D) >= (end(Bin,D)/2)))),
   100*max(L,R) =< (100+Ratio)*min(L,R))).
```

express particular weight constraints in an admissible packing.

The complete PKML library including common sense rules dealing with the weight of objects and the surface contact of stacked items, is given in [14]. With these rules, Proposition 3 entails:

Proposition 4. *PKML models containing lists of at most l elements generate constraint programs of size $O(l^4)$ in presence of both alternative shapes and assemblies of boxes, $O(l^3)$ in presence of only one of them, and $O(l^2)$ in presence of single box shapes only.*

Business patterns can also be used in PKML to express knowledge about some predefined (partial) solutions to packing problems. Such patterns are used in the industry, for instance for filling pallets, or trucks, with maximum stability according to some predefined solutions. Stability conditions can be expressed with non-guillotine or non-visibility constraints [20], however packing patterns provide a pragmatic and complementary approach to these important requirements. In PKML, packing patterns can be defined as records containing a list of item shapes given with the coordinates of their origin, and bounds on their weight.

4.4   Performance Evaluation

We report here the performances of the Rules2CP compiler and of the generated constraint program, on Korf's benchmarks of optimal rectangle packing problems [7]. These problems consists in finding the smallest rectangle containing n squares of sizes $S_i = i$ for $1 \leq i \leq n$. In [21] Simonis and O'Sullivan proposed a constraint program implemented in SICStus Prolog which improved best known runtimes up to a factor of 300.

Their search strategy decomposes the optimisation problem in two subproblems. First, the different non symmetric bounding rectangle candidates are enumerated in ascending order of areas. Then, for each bounding rectangle candidate, the N squares packing satisfaction problem is solved with a search strategy based on interval splitting, working together with the disjoint2 and cumulative global constraints. The strategy places the N squares ordered by decreasing sizes. It first splits the domain of x coordinates into intervals, before fixing these coordinates by dichotomy. The process is then repeated for the y coordinates.

Table 1. Optimal Rectangle Packing programs runtimes in seconds

N	R2CP compilation	Rules2CP	Original
18	0.266	13	6
19	0.310	11	5
20	0.320	20	10
21	0.342	76	36
22	0.369	364	197
23	0.404	2076	1150
24	0.443	5230	1847
25	0.509	52909	17807

Table 1 compares the computation time in seconds obtained in Rules2CP with their original program in SICStus-Prolog . The SICStus-Prolog program generated from the Rules2CP model with dynamic search explores exactly the same search space and is slower by a factor less than 3, due to the interpretation overhead for the dynamic predicates. In all these examples, the compilation times are below one second.

5   Related Work

5.1   Comparison with OPL, Zinc and Essence

Rules2CP differs from OPL [1], Zinc [3,4] and Essence [5] modelling languages in several aspects, among which: the naming of rules, the restriction to simple data structures of records and enumerated lists, the absence of recursion, the declarative specification of heuristics as preference orderings, and the absence of program annotations.

This trade-off for ease of use was motivated by our search for a declarative modelling language with no complicated programming constructs. We have shown that the declarations and rules of Rules2CP allow the user to give names to data and knowledge rules without complicated variable scope. A simple module system is used in Rules2CP to avoid name clashes. The simplicity of these design choices is reflected in the obtainment of a complexity bound on the size of the constraint programs generated from Rules2CP models (Prop. 3). Moreover, the partial evaluation mechanism used in the rewriting process eliminates at compile-time the overhead due to the simplicity of our data and control structures.

Interestingly, we have shown that complex search strategies can be expressed *declaratively* in Rules2CP, by specifying *decision variables* and *branching formulas*, as well as both static and dynamic choice heuristics as *preference orderings* on variables and values. These specifications use all the power of the language to define heuristic criteria. This is currently not expressible in Zinc and Essence, and can be achieved in OPL in aless declarative manner by programming. On the other hand, we have not considered the compilation of Rules2CP to other solvers such as local search, or mixed integer linear programs, as has been done for OPL and Zinc systems.

5.2 Comparison with Constraint Logic Programming

As a modelling language, Rules2CP is a constraint logic programming language, but not in the formal sense of the CLP scheme of Jaffar and Lassez [22]. Rules2CP models can be compiled to CLP(FD) programs of potentially exponential size. Note that the converse translation of Prolog programs into Rules2CP models is not possible (apart from an arithmetic encoding) because of the absence of recursion and of general list constructors in Rules2CP. Furthermore, free variables are not allowed in the right hand side of Rules2CP rules. Instead of the local scope mechanism used for the free variables in CLP rules, a global scope mechanism in used for the free variables in Rules2CP declarations. This global scope mechanism has no counterpart in the CLP scheme which makes it often necessary to pass the list of all variables as arguments to CLP predicates.

5.3 Comparison with Business Rules

Rules2CP is an attempt to use the business rules knowledge representation paradigm for constraint programming. Business rules are very popular in the industry because they provide a declarative mean for expressing expertise knowledge. Business rules should describe independent pieces of knowledge, and should be independent from a particular procedural interpretation by a rule engine [6]. Rules2CP realizes this aim in the context of combinatorial optimisation problems, by tranforming business rules into efficient programs using completely different representations. Rules2CP rules are not general condition-action rules, also called production rules in the expert system community, but *logical rules* with only one head and no imperative actions. Bounded quantifiers are used to represent complex conditions. Such conditions can also be expressed in many production rules systems, but here they are used at compile-time to setup a constraint satisfaction problem, instead of at run-time to match patterns in a database of facts.

5.4   Comparison with Term Rewriting Systems Tools

The compilation of Rules2CP models to constraint programs is defined and implemented by a term rewriting system. The properties of confluence and termination of this process have been shown using term rewriting theory. There are several term rewriting system tools available that could be directly used for the implementation of the Rules2CP compiler. For instance, in the context of target constraint solvers in Java, such as e.g. Choco, and for Java programming environments in which Rules2CP data structures may be defined by Java objects, the term rewriting system TOM [23] provides a pattern matching compiler for programming term transformations defined by rules. This would make of TOM an ideal system for implementing a Rules2CP compiler to Java, through a direct translation of $\rightarrow$ rules into TOM pattern matching expressions.

6   Conclusion

The Rules2CP language is a rule-based modelling language for constraint programming. It has been designed to allow application experts express knowledge, common sense and industrial requirements about combinatorial optimisation problems with rules (using appropriate editors). Rules2CP rules are declarative and can be easily introduced, checked and modified one by one, independently from their particular interpretation by a rule engine.

Search trees can also be specified declaratively in Rules2CP with logical formulae, and search heuristics can be defined as preference orderings on variables, values, conjunctive and disjunctive formulae, using pattern matching on rule names. This is in contrast with other modelling languages for which search strategies still need be programmed. We have shown that search strategies for scheduling can be easily expressed in Rules2CP in this manner, as well as the search strategies of Simonis and O'Sullivan [21] for solving Korf's optimal rectangle packing problem [7], with a constant overhead factor in the generated code.

The PKML library dedicated to bin packing and bin design problems used in these experiments, can deal in addition with extra requirements about weights, oversizes, equilibrium constraints, and specific packing business rules. Furthermore, a large subset of PKML has been shown in [16] to be efficiently compilable with indexicals within the geometrical kernel of the global constraint geost.

The transformation of Rules2CP models into constraint programs has been described here by a term rewriting system with partial evaluation. The confluence of these transformations has been shown, together with a complexity bound on the size of the generated program. The obtainment of such a complexity result reflects the simplicity of our design choices for Rules2CP, such as the absence of recursion and of general list constructor for instance. This complexity bound shows however a potential exponential blow-up in the size of the generated constraints. In such cases, the dynamical expansion strategy can be used.

As for future work, several issues have not been discussed in this paper. Rules2CP is currently untyped. One difficulty in typing Rules2CP models lies in the coercions between expressions and formulae used in reification and involving a subtyping relation between booleans and integers [24]. More experiments are also needed to evaluate the module system of Rules2CP and its capability to develop libraries of models that can be reused in a hierarchy of models and for special purpose applications. Finally, the specification of search strategies in Rules2CP needs be explored more systematically, and could also be evaluated with adaptive strategies in which the dynamic criteria depend on execution profiling criteria.

Acknowledgements. This work is supported by the European FP7 Strep project Net-WMS. We gratefully acknowledge Helmut Simonis for providing us with his program, and all the partners of this project, as well as Thierry Martinez and Sylvain Soliman, for numerous discussions on this topic.

References

1. Van Hentenryck, P.: The OPL Optimization programming Language. MIT Press, Cambridge (1999)
2. Hentenryck, P.V., Perron, L., Puget, J.F.: Search and strategies in opl. ACM Transactions on Compututational Logic 1, 285–320 (2000)
3. Rafeh, R., de la Banda, M.G., Marriott, K., Wallace, M.: From Zinc to design model. In: Hanus, M. (ed.) PADL 2007. LNCS, vol. 4354, pp. 215–229. Springer, Heidelberg (2006)
4. de la Banda, M.G., Marriott, K., Rafeh, R., Wallace, M.: The modelling language Zinc. In: Benhamou, F. (ed.) CP 2006. LNCS, vol. 4204, pp. 700–705. Springer, Heidelberg (2006)
5. Frisch, A.M., Harvey, W., Jefferson, C., Martinez-Hernandez, B., Miguel, I.: Essence: A constraint language for specifying combinatorial problems. Constraints 13, 268–306 (2008)
6. Group, B.R.: The business rules manifesto Business Rules Group (2003), http://www.businessrulesgroup.org/brmanifesto.htm
7. Korf, R.E.: Optimal rectangle packing: New results. In: ICAPS, pp. 142–149 (2004)
8. Haemmerlé, R., Fages, F.: Modules for prolog revisited. In: Etalle, S., Truszczyński, M. (eds.) ICLP 2006. LNCS, vol. 4079, pp. 41–55. Springer, Heidelberg (2006)
9. Van Hentenryck, P.: Constraint satisfaction in Logic Programming. MIT Press, Cambridge (1989)
10. Huang, J., Darwiche, A.: The language of search. Journal of Artificial Intelligence Research 29, 191–219 (2007)
11. Apt, K., Wallace, M.: Constraint Logic Programming using Eclipse. Cambridge University Press, Cambridge (2006)
12. Carlsson, M., et al.: SICStus Prolog User's Manual. Swedish Institute of Computer Science, 4th edn. (2007), ISBN 91-630-3648-7

13. Fages, F., Soliman, S., Coolen, R.: CLPGUI: a generic graphical user interface for constraint logic programming. Journal of Constraints, Special Issue on User-Interaction in Constraint Satisfaction 9, 241–262 (2004)
14. Fages, F., Martin, J.: From rules to constraint programs with the Rules2CP modelling language. INRIA Research Report RR-6495, Institut National de Recherche en Informatique (2008)
15. Rosen, B.: Tree-manipulating systems and Church-Rosser theorems. Journal of the ACM 20, 160–187 (1973)
16. Carlsson, M., Beldiceanu, N., Martin, J.: A geometric constraint over k-dimensional objects and shapes subject to business rules. In: Stuckey, P.J. (ed.) CP 2008. LNCS, vol. 5202, pp. 220–234. Springer, Heidelberg (2008)
17. Beldiceanu, N., Carlsson, M., Poder, E., Sadek, R., Truchet, C.: A generic geometrical constraint kernel in space and time for handling polymorphic k-dimensional objects. In: Bessière, C. (ed.) CP 2007. LNCS, vol. 4741, pp. 180–194. Springer, Heidelberg (2007); SICS Technical Report T2007:08, http://www.sics.se/libindex.html
18. Allen, J.: Time and time again: The many ways to represent time. International Journal of Intelligent System 6 (1991)
19. Randell, D., Cui, Z., Cohn, A.: A spatial logic based on regions and connection. In: Nebel, B., Rich, C., Swartout, W.R. (eds.) Proc. of 2nd International Conference on Knowledge Representation and reasoning KR 1992, pp. 165–176. Morgan Kaufmann, San Francisco (1992)
20. Carpenter, H., Dowsland, W.: Practical consideration of the pallet loading problem. Journal of the Operations Research Society 36, 489–497 (1985)
21. Simonis, H., O'Sullivan, B.: Using global constraints for rectangle packing. In: Proceedings of the first Workshop on Bin Packing and Placement Constraints BPPC 2008, associated to CPAIOR 2008 (2008)
22. Jaffar, J., Lassez, J.L.: Constraint logic programming. In: Proceedings of the 14th ACM Symposium on Principles of Programming Languages, pp. 111–119. ACM, Munich (1987)
23. Balland, E., Brauner, P., Kopetz, R., Moreau, P.E., Reilles, A.: Tom: Piggybacking rewriting on java. In: Baader, F. (ed.) RTA 2007. LNCS, vol. 4533, pp. 36–47. Springer, Heidelberg (2007)
24. Fages, F., Coquery, E.: Typing constraint logic programs. Journal of Theory and Practice of Logic Programming 1, 751–777 (2001)

Combining Symmetry Breaking and Global Constraints*

George Katsirelos, Nina Narodytska, and Toby Walsh

NICTA and UNSW, Sydney, Australia

Abstract. We propose a new family of constraints which combine together lexicographical ordering constraints for symmetry breaking with other common global constraints. We give a general purpose propagator for this family of constraints, and show how to improve its complexity by exploiting properties of the included global constraints.

1   Introduction

The way that a problem is modeled is critically important to the success of constraint programming. Two important aspects of modeling are symmetry and global constraints. A common and effective method of dealing with symmetry is to introduce constraints which eliminate some or all of the symmetric solutions [1]. Such symmetry breaking constraints are usually considered separately to other (global) constraints in a problem. However, the interaction between problem and symmetry breaking constraints can often have a significant impact on search. For instance, the interaction between problem and symmetry breaking constraints gives an exponential reduction in the search required to solve certain pigeonhole problems [2]. In this paper, we consider even tighter links between problem and symmetry breaking constraints. We introduce a family of global constraints which combine together a common type of symmetry breaking constraint with a range of common problem constraints. This family of global constraints is useful for modeling scheduling, rostering and other problems.

Our focus here is on matrix models [3]. Matrix models are constraint programs containing matrices of decision variables on which common patterns of constraints are posted. For example, in a rostering problem, we might have a matrix of decision variables with the rows representing different employees and the columns representing different shifts. A problem constraint might be posted along each row to ensure no one works too many night shifts in any 7 day period, and along each column to ensure sufficient employees work each shift. A common type of symmetry on such matrix models is row interchangeability [4]. Returning to our rostering example, rows representing equally skilled employees might be interchangeable. An effective method to break such symmetry is to order lexicographically the rows of the matrix[4]. To increase the propagation between such symmetry breaking and problem constraints, we

---

* NICTA is funded by the Australian Government as represented by the Department of Broadband, Communications and the Digital Economy and the Australian Research Council through the ICT Centre of Excellence program.

A. Oddi, F. Fages, and F. Rossi (Eds.): CSCLP 2008, LNAI 5655, pp. 84–98, 2009.

consider compositions of lexicographical ordering and problem constraints. We conjecture that the additional pruning achieved by combining together symmetry breaking and problem constraints will justify the additional cost of propagation. In support of this, we present a simple problem where it gives a super-polynomial reduction in search. We also implement these new propagators and run them on benchmark nurse scheduling problems. Experimental results show that propagating of a combination of symmetry breaking and global constraints reduces the search space significantly and improves run time for most of the benchmarks.

2 Background

A constraint satisfaction problem (CSP) P consists of a set of variables $\mathcal{X} = \{X[i]\}$, $i = 1, \ldots, n$ each of which has a finite domain $D(X[i])$, and a set of constraints $\mathcal{C}$. We use capital letters for variables (e.g. $X[i]$ or $Y[i]$), lower case for values (e.g. v or v_i) and write $\boldsymbol{X}$ for the sequence of variables, $X[1]$ to $X[n]$. A constraint $C \in \mathcal{C}$ has a *scope*, denoted $scope(C) \subseteq \mathcal{X}$ and allows a subset of the possible assignments to the variables $scope(C)$, called *solutions* or *supports* of C. A constraint is *domain consistent (DC)* iff for each variable $X[i]$, every value in the domain of X_i belongs to a support. A solution of a CSP P is an assignment of one value to each variable such that all constraints are satisfied. A matrix model of a CSP is one in which there is one (or more) matrices of decision variables. For instance, in a rostering problem, one dimension might represent different employees and the other dimension might represent days of the week.

A common way to solve a CSP is with backtracking search. In each node of the search tree, a *decision* restricts the domain of a variable and the solver infers the effects of that decision by invoking a *propagator* for each constraint. A propagator for a constraint C is an algorithm which takes as input the domains of the variables in $scope(C)$ and returns *restrictions* of these domains. We say the a propagator enforces *domain consistency (DC)* on a constraint C iff an invocation of the propagator ensures that the constraint C is domain consistent.

A *global constraint* is a constraint in which the number of variables is not fixed. Many common and useful global constraints have been proposed. We introduce here the global constraints used in this paper. The global lexicographical ordering constraint $\text{LEX}(\boldsymbol{X}, \boldsymbol{Y})$ is recursively defined to hold iff $X[1] < Y[1]$, or $X[1] = Y[1]$ and $\text{LEX}([X[2], \ldots, X[n]], [Y[2], \ldots, Y[n]])$ [5]. This constraint is used to break symmetries between vectors of variables. The global sequence constraint $\text{SEQUENCE}(l, u, k, \boldsymbol{X}, V)$ holds iff $l \leq |\{i \mid X[i] \in V, j \leq i < j + k\}| \leq u$ for each $1 \leq j < n - k$ [6]. The regular language constraint $\text{REGULAR}(\mathcal{A}, \boldsymbol{X})$ holds iff $X[1]$ to $X[n]$ takes a sequence of values accepted by the deterministic finite automaton $\mathcal{A}$ [7]. The last two constraints are useful in modeling rostering and scheduling problems.

3 The C&LEX Constraint

Two common patterns in many matrix models are that rows of the matrix are interchangeable, and that a global constraint C is applied to each row. To break such row symmetry, we can post constraints that lexicographically order rows [4]. To improve

propagation between the symmetry breaking and problem constraints, we propose the
C&LEX(X, Y, C) constraint. This holds iff $C(X)$, $C(Y)$ and LEX(X, Y) all simultaneously hold. To illustrate the potential value of such a C&LEX constraint, we give a
simple example where it reduces search super-polynomially.

Example 1. Let M be a $n \times 3$ matrix in which all rows are interchangeable. Suppose
that $C(X, Y, Z)$ ensures $Y = X + Z$, and that variable domains are as follows:

$$M = \begin{pmatrix} \{1, \ldots, n-1\} & \{n+1, \ldots, 2n-1\} & n \\ \{1, \ldots, n-1\} & \{n, \ldots, 2n-2\} & n-1 \\ \ldots & \ldots & \ldots \\ \{1, \ldots, n-1\} & \{3, \ldots, n+1\} & 2 \\ \{1, \ldots, n-1\} & \{2, \ldots, n\} & 1 \end{pmatrix}.$$

We assume that the branching heuristic instantiates variables top down and left to
right, trying the minimum value first. We also assume we enforce DC on posted constraints. If we model the problem with C&LEX constraints, we solve it without search.
On the other hand, if we model the problem with separate LEX and C constraints, we
explore an exponential sized search tree before detecting inconsistency using the mentioned branching heuristic and a super-polynomial sized tree with any k-way branching
heuristic.

3.1 Propagating C&LEX

We now show how, given a (polynomial time) propagator for the constraint C, we
can build a (polynomial time) propagator for C&LEX. The propagator is inspired by
the DC filtering algorithm for the LEXCHAIN constraint proposed by Carlsson and
Beldiceanu [8]. The LEXCHAIN constraint ensures that rows of the matrix M are
lexicographically ordered. If the LEXCHAIN constraint is posted on two rows then
LEXCHAIN is equivalent to the C&LEX(X, Y, True) constraint. However, unlike [8],
we can propagate here a conjunction of the LEX constraint and arbitrary global constraints C. The propagator for the C&LEX constraint is based on the following result
which decomposes propagation into two simpler problems.

Proposition 1. *Let X_l be the lexicographically smallest solution of $C(X)$, Y_u be the
lexicographically greatest solution of $C(Y)$, and LEX(X_l, Y_u). Then enforcing DC
on C&LEX(X, Y, C) is equivalent to enforcing DC on C&LEX(X, Y_u, C) and on
C&LEX(X_l, Y, C).*

Proof. Suppose C&LEX(X_l, Y, C) is DC. We are looking for support for $Y_k = v$, where Y_k is an arbitrary variable in Y. Let Y' be a support for $Y_k = v$
in C&LEX(X_l, Y, C). Such a support exists because C&LEX(X_l, Y, C) is DC.
C&LEX(X_l, Y, C) ensures that Y' is a solution of $C(Y)$ and LEX(X_l, Y'). Consequently, X_l and Y' are a solution of C&LEX(X, Y, C). Similarly, we can find a
support for $X_k = v$, where X_k is an arbitrary variable in X.   □

Thus, we will build a propagator for C&LEX that constructs the lexicographically smallest (greatest) solution of $C(X)$ ($C(Y)$) and then uses two simplified C&LEX constraints
in which the first (second) sequence of variables is replaced by the appropriate bound.

Finding the lexicographically smallest solution. We first show how to find the lexi-cographically smallest solution of a constraint. We denote this algorithm $C_{min}(L, X)$. A dual method is used to find the lexicographically greatest solution. We use a greedy algorithm that scans through X and extends the partial solution by selecting the small-est value from the domain of $X[i]$ at ith step (line 6). To ensure that the selection at the next step will never lead to a failure, the algorithm enforces DC after each value selection (line 7). Algorithm 1 gives the pseudo-code for the $C_{min}(L, X)$ algorithm. The time complexity of Algorithm 1 is $O(nc + nd)$, where d is the total number of values in the domains of variables X and c is the (polynomial) cost of enforcing DC on C.

Algorithm 1. $C_{min}(L, X)$

```
1: procedure C_min(L : out, X : in)
2:    if (DC(C(X)) == fail) then
3:       return false;
4:    Y = Copy(X);
5:    for i = 1 to n do
6:       Y[i] = L[i] = min(D(Y[i]));
7:       DC(C(Y));
8:    return true;
```

Proposition 2. *Let $C(X)$ be a global constraint. Algorithm 1 returns the lexicograph-ically smallest solution of the global constraint C if such a solution exists.*

Proof. First we prove that if there is a solution to $C(X)$ then Algorithm 1 returns a solution. Second, we prove that the solution returned is the lexicographically smallest solution.

1. If $C(X)$ does not have a solution then Algorithm 1 fails at line 3. Otherwise $C(X)$ has a solution. Since $DC(C(X))$ leaves only consistent values, any value of $X[1]$ can be extended to a solution of $C(X)$ and Algorithm 1 selects $L[1]$ to be the minimum value of $X[1]$. Suppose Algorithm 1 performed $i - 1$ steps and the partial solution is $[L[1], \ldots, L[i-1]]$. All values left in the domains of at $X[i], \ldots, X[n]$ are consistent with the partial solution $[L[1], \ldots, L[i-1]]$. Consequently, any value that is in the domain of $X[i]$ is consistent with $[L[1], \ldots, L[i-1]]$ and can be extended to a solution of $C(X)$. The algorithm assigns $L[i]$ to the minimum value of $X[i]$. Moving forward to the end of the sequence, the algorithm finds a solution to $C(X)$.
2. By contradiction. Let L' be the lexicographically smallest solution of $C(X)$ and L be the solution returned by Algorithm 1. Let i be the first position where L' and L differ so that $L'[i] < L[i]$, $L'[k] = L[k]$, $k = 1, \ldots, i - 1$. Consider ith step of Algorithm 1. As $DC(C(X))$ is correct, all values of $X[i]$ consistent with $[L[1], \ldots, L[i-1]]$ are in the domain of $X[i]$. The algorithm selects $L[i]$ to be equal to $min(D(X[i]))$. Therefore, $[L[1], \ldots, L[i]]$ is the lexicographically smallest pre-fix of length i for a solution of $C(X)$. Hence, there is no solution of $C(X)$ with prefix $[L'[1], \ldots, L'[i]] \leq_{lex} [L[1], \ldots, L[i]]$. This leads to a contradiction.

A filtering algorithm for the $C\&LEX_{lb}(L, X, C)$ **constraint.** The propagation algorithm for the $C\&LEX_{lb}(L, X, C)$ constraint finds all possible supports that are greater than or equal to the lower bound L and marks the values that occur in these supports. Algorithm 2 gives the pseudo-code for the propagator for $C\&LEX_{lb}$. The algorithm uses the auxiliary routine $MarkConsistentValues(C, X, X')$. This finds all values in domains of X' that satisfy $C(X')$ and marks corresponding values in X. The time complexity of the $MarkConsistentValues(C, X, X')$ procedure is $O(nd + c)$. The total time complexity of the propagator for the $C\&LEX_{lb}$ filtering algorithm is $O(n(nd+c))$. A dual algorithm to $C\&LEX_{lb}$ is $C\&LEX_{ub}(X, U, C)$ that finds all possible supports that are less than or equal to the upper bound U and marks the values that occur in these supports.

Algorithm 2. $C\&LEX_{lb}(L, X, C)$

1: **procedure** $C\&LEX_{lb}(L : out, X : out, C : in)$
2:      **if** $(DC(C(X)) == fail)$ **then**
3:          return $false$;
4:      $LX = X$;
5:      **for** $i = 1$ **to** n **do**
6:          $D(LX[i]) = \{v_j | v_j \in D(LX[i]) \text{ and } L[i] < v_j\}$;
7:          $MarkConsistentValue(C, X, LX)$;
8:          **if** $L[i] \notin D(X[i])$ **then**
9:              break;
10:         **else**
11:             $LX[i] = L[i]$;
12:     **if** $(i == n)$ **then**
13:         $MarkConsistentValues(C, X, L)$;
14:     **for** $i = 1$ **to** n **do**
15:         $Prune(\{v_j \in D(X[i]) | unmarked(v_j)\})$;

Algorithm 3. Mark consistent values

1: **procedure** $MarkConsistentValues(C : in, X : out, X' : in)$
2:      $Z = Copy(X')$;
3:      $DC(C(Z))$;
4:      **for** $i = 1$ **to** n **do**
5:          $Mark\{v_j | v_j \in D(X[i]) \text{ and } v_j \in D(Z[i])\}$;

We also need to prove that Algorithm 2 enforces domain consistency on the $C\&LEX_{lb}(L, X, C)$ constraint. A dual proof holds for $C\&LEX_{ub}$.

Proposition 3. *Algorithm 2 enforces DC on the* $C\&LEX_{lb}(L, X, C)$ *constraint.*

Proof. We first show that if a value v was not pruned from the domain of $X[p]$ (or marked) then it does have a support for $C\&LEX_{lb}(L, X, C)$. We then show that if a value v was pruned from the domain of $X[p]$ (or not marked) then it does not have a support.

1. Algorithm 2 marks values in two lines 7 and 13. Suppose at step i the algorithm marks value $v \in D(X[p])$ at line 7. At this point we have that $LX[k] = L[k]$, $k = 1, \ldots, i-1, L[i] < LX[i]$. After enforcing DC on $C(LX)$, the value v is left

in the domain of $LX[p]$. Consequently, there exists a support for $X[p] = v$, starting with $[L[1], \ldots, L[i-1], v', \ldots]$, $v' \in D(LX[i])$, that is strictly greater than L. Marking at line 13 covers the case where L is a solution of $C(X)$.

2. By contradiction. Suppose that value $v \in D(X[p])$ was not marked by Algorithm 2 but it has a support X' such that $L \leq_{lex} X'$. Let i be the first position where $L[i] < X[i]$ and $L[k] = X[k]$, $k = 1, \ldots, i-1$. We consider three disjoint cases:

   – The case that no such i exists. Then L is a support for value $v \in D(X[p])$. Hence, value v has to be marked at line 13. This leads to a contradiction.

   – The case that $i \leq n$ and $p < i$. Note that in this case v equals $L[p]$. Consider Algorithm 2 at step i. At this point we have $L[k] = LX[k]$, $k = 1, \ldots, i-1$. After enforcing DC on $C(LX)$ (line 7), values $X'[i]$, $i = 1, \ldots, n$ are left in the domain of LX, because $L[i] < X'[i]$, $L[k] = X'[k]$, $k = 1, \ldots, i-1$. Hence, value $v \in X'[p]$ will be marked at line 7. This leads to a contradiction.

   – The case that $i \leq n$ and $i \leq p$. Consider Algorithm 2 at step i. At this point we have $L[k] = LX[k]$, $k = 1, \ldots, i-1$. Moreover, value $X'[i]$ has to be in the domain of $LX[i]$, because value $X'[i]$ is greater than $L[i]$ and is consistent with the partial assignment $[L[1], \ldots, L[i-1]]$. Domains of variables LX contain all values that have supports starting with $[L[1], \ldots, L[i-1]]$ and are strictly greater than L. Consequently, they contain $X'[i]$, $i = 1, \ldots, n$ and the algorithm marks v at line 7. This leads to a contradiction.

$\square$

A filtering algorithm for the C&LEX(X, Y, C). Algorithm 4 enforces domain consistency on the C&LEX(X, Y, C) constraint. Following Proposition 1, Algorithm 4 finds the lexicographically smallest (greatest) solutions for $C(X)$ $(C(Y))$ and runs a relaxed version of C&LEX for each row. Algorithm 4 gives the pseudo-code for the propagator for the C&LEX(X, Y, C) constraint.

Algorithm 4. C&LEX(X, Y, C)

```
1: procedure C&LEX(X : out, Y : out, C : in)
2:     if (C_min(X_l, X) == fail) or (C_max(Y, Y_u)) == fail) then
3:         return false;
4:     if (X_l >_lex Y_u) then
5:         return false;
6:     C&LEX_lb(X_l, Y, C);
7:     C&LEX_ub(X, Y_u, C);
```

Proposition 4. *Algorithm 4 enforces DC on the C&LEX(X, Y, C) constraint.*

Proof. Correctness of the algorithm follows from correctness of the decomposition (Proposition 1). However, we need to consider the case where $X_l >_{lex} Y_u$, prove correctness of the C&LEX$_{lb}$ and C&LEX$_{ub}$ algorithms and prove that the algorithm only needs to run once.

If $X_l >_{lex} Y_u$ then C&LEX(X, Y, C) does not have a solution and Algorithm 4 fails at line 5. Otherwise, we notice that if $X_l \leq_{lex} Y_u$ then X_l and Y_u is a solution of C&LEX(X, Y, C), because X_l is a solution of $C(X)$, Y_u is a solution of $C(Y)$ and $X_l \leq_{lex} Y_u$. Consequently, invocation of the simplified version of C&LEX at lines 6 and 7 cannot change X_l and Y_u. $\square$

Example 2. We consider how Algorithm 4 works on the first two rows C&LEX constraint from Example 1. Let n equal 5. In this case domains of the first two rows of variables are $\begin{pmatrix} M[1] \\ M[2] \end{pmatrix} = \begin{pmatrix} [1, 2, 3, 4] \ [6, 7, 8, 9] \ 5 \\ [1, 2, 3, 4] \ [5, 6, 7, 8] \ 4 \end{pmatrix}$.

Suppose the solver branches on $X[1] = 1$. Algorithm 4 finds the lexicographically smallest and greatest solutions of $M[1]$ and $M[2]$ using Algorithm 1(line 2). These solutions are $[1, 6, 5]$ and $[4, 8, 4]$ respectively . Then enforces DC on C&LEX$_{lb}([1, 6, 5], M[2], C)$ in the following way:

1. copies $M[2]$ to $\boldsymbol{LX}$
2. marks all values that have a support starting with a value greater than 1 (that is 2, 3 and 4). There are three supports that satisfy this condition, namely, $[2, 6, 4]$, $[3, 7, 4]$ and $[4, 8, 4]$. Checks conditions at line 8 and assigns $LX[1]$ to 1. Then it moves to the next iteration.
3. marks all values that have a support starting with a prefix greater than $[1, 6]$. There are no such values. Checks conditions at line 8 and assigns $LX[2]$ to 6. Then it moves to the next iteration.
4. marks all values that have a support starting with a prefix greater than $[1, 6, 5]$. There are no such values. Checks conditions at line 8 and stops the marking part.
5. removes unmarked values: value 1 from $X[2]$ and value 5 from $Y[2]$.

Finally, it enforces DC on C&LEX$_{ub}(M[1], [4, 8, 4], C)$. This sets $M[1]$ to $[1, 6, 5]$, because the solver branched on $X[1] = 1$ and $[1, 6, 5]$ is the only possible support for this assignment.

The time complexity of the general algorithm is more expensive than the decomposition into individual constraints $C(\boldsymbol{X})$, $C(\boldsymbol{Y})$ and LEX$(\boldsymbol{X}, \boldsymbol{Y})$ by a linear factor. The general algorithm is not incremental, but its performance can be improved by detecting entailment. If $\boldsymbol{X_u} < \boldsymbol{Y_l}$ then the LEX constraint is entailed and C&LEX can be decomposed into two constraints $C(\boldsymbol{X})$ and $C(\boldsymbol{Y})$. Similarly, we can improve the complexity by detecting when $C(\boldsymbol{X})$ and $C(\boldsymbol{Y})$ are entailed. As we show in the next sections, the time complexity of the propagator for the C&LEX$(\boldsymbol{X}, \boldsymbol{Y}, C)$ constraint can also be improved by making it incremental for many common constraints C by exploiting properties of C. Note also that Algorithm 4 easily extends to the case that different global constraints are applied to $\boldsymbol{X}$ and $\boldsymbol{Y}$.

3.2   The C&LEX$(\boldsymbol{X}, \boldsymbol{Y}, $ SEQUENCE$)$ Constraint

In this section we consider the case of a conjunction of the LEX constraint with two SEQUENCE constraints. First we assume that variables $\boldsymbol{X}$ and $\boldsymbol{Y}$ are Boolean variables. Later we will show how to extend this to the general case. In the Boolean case, we can exploit properties of the filtering algorithm for the SEQUENCE constraint ($HPRS$) proposed in   [9]. The core of the $HPRS$ algorithms is the CheckConsistency procedure that detects inconsistency if the SEQUENCE constraint is unsatisfiable and returns the lexicographically smallest solution otherwise. The $HPRS$ algorithm runs CheckConsistency for each variable-value pair $X_i = v_j$. If CheckConsistency detects a failure, then value v_j can be pruned from $D(X_i)$, otherwise CheckConsistency returns the lexicographically smallest support for $X_i = v_j$.

As was shown in [9], the algorithm can be modified so that CheckConsistency returns the lexicographically greatest support. Both versions of the algorithm are useful for us. We will use the min subscript for the first version of the algorithm, and the max subscript for the second.

Due to these properties of the $HPRS$ algorithm, a propagator for the C&LEX $(X, Y_u, \text{SEQUENCE})_{lb}$, denoted $HPRS'_{min}(X, Y_u)$, is a slight modification of $HPRS_{min}$, which checks that the lexicographically smallest support for $X_i = v_j$ returned by the CheckConsistency$_{min}$ procedure is lexicographically smaller than or equal to Y_u. To find the lexicographically greatest solution, Y_u, of the SEQUENCE(Y) constraint, we run CheckConsistency$_{max}$ on variables Y. Dual reasoning is applied to the C&LEX$(X_l, Y, \text{SEQUENCE})_{ub}$ constraint. Algorithms 5 shows pseudo code for DC propagator for the C&LEX$(X, Y, \text{SEQUENCE})$ constraint.

Algorithm 5. C&LEX$(X, Y, \text{SEQUENCE})$

```
1: procedure C&LEX(X : out, Y : out, SEQUENCE (l, u, k) : in)
2:     if ¬(CheckConsistency_min(X_l, X)) or ¬(CheckConsistency_max(Y, Y_u)) then
3:         return false;
4:     if (X_l >_lex Y_u) then
5:         return false;
6:     HPRS'_max(X_l, Y, SEQUENCE (l, u, k));
7:     HPRS'_min(X, Y_u, SEQUENCE (l, u, k));
```

$HPRS'_{min}$ and $HPRS'_{max}$ are incremental algorithms, therefore the total time complexity of Algorithm 5 is equal to the complexity of the $HPRS$ algorithm, which is $O(n^3)$ down a branch of the search tree. Correctness of Algorithm 5 follows from Proposition 1 and correctness of the $HPRS$ algorithm.

Example 3. Consider the SEQUENCE$(2, 2, 3, [X[1], X[2], X[3], X[4]])$ and SEQUENCE$(2, 2, 3, [Y[1], Y[2], Y[3], Y[4]])$ constraints. The domains of the variables are $X = [\{0, 1\}, \{1\}, \{0, 1\}, \{0, 1\}]$ and $Y = [\{0, 1\}, \{0, 1\}, \{1\}, \{0, 1\}]$. Note that each of the two SEQUENCE and the LEX(X, Y) constraints are domain consistent.

The C&LEX$(X, Y, \text{SEQUENCE})$ constraint fixes variables X to $[0, 1, 1, 0]$. The lexicographically greatest solution for the SEQUENCE(Y) is $[1, 0, 1, 0]$, while the lexicographically smallest support for $X[1] = 1$ is $[1, 1, 0, 1]$. Therefore, the value 1 will be pruned from the domain of $X[1]$. For the same reason, the value 0 will be pruned from $X[3]$ and the value 1 will be pruned from $X[4]$.

Consider the general case, where X and Y are finite domain variables. We can channel the variables X, Y into Boolean variables b_X, b_Y and post SEQUENCE(b_X), SEQUENCE(b_Y), which does not hinder propagation. Unfortunately, we cannot post the LEX constraint on the Boolean variables b_X and b_Y, because some solutions will be lost. For example, suppose we have SEQUENCE$(X, 0, 1, 2, \{2, 3\})$ and SEQUENCE$(Y, 0, 1, 2, \{2, 3\})$ constraints. Let $X = [2, 0, 2]$ and $Y = [3, 0, 0]$ be solutions of these constraints. The corresponding Boolean variables are $b_X = [1, 0, 1]$ and $b_Y = [1, 0, 0]$. Clearly $X <_{lex} Y$, but $b_X >_{lex} b_Y$. Therefore, the LEX constraint can be enforced only on the original variables.

The problem is that the $HPRS$ algorithm returns the lexicographically smallest solution on Boolean variables. As the example above shows, lexicographical comparison between Boolean solutions of SEQUENCES b_X and b_Y is not sound with respect to the original variables. Therefore, given a solution of SEQUENCE(b_X), we need to find the corresponding lexicographically smallest solution of SEQUENCE(X). We observe that if we restrict ourselves to a special case of SEQUENCE(l, u, k, v, X) where $\max(D \setminus v) < \min(v)$ then this problem can be solved in linear time as follows. Let b_X be a solution for SEQUENCE(b_X). Then the corresponding lexicographically smallest solution X for SEQUENCE(X) is $X[i] = min(v \cap D(X[i]))$ if $b_X[i] = 1$ and $X[i] = min(D(X[i]))$ otherwise. In a similar way we can find the corresponding lexicographically greatest solution. A slight modification to Algorithm 5 is needed in this case. Whenever we need to check whether b_X is smaller than or equal to b_Y, we transform b_X to the corresponding lexicographically smallest solution, b_Y to the corresponding lexicographically greatest solution and perform the comparison.

3.3   The C&LEX(X, Y, REGULAR) Constraint

With the REGULAR($\mathcal{A}, X$) constraint, we will show that we can build a propagator for C&LEX which takes just $O(nT)$ time, compared to $O(n^2T)$ for our general purpose propagator, where d is the maximum domain size and T is the number of transitions of the automaton $\mathcal{A}$. We will use the following example to illustrate results in this section.

Example 4. Consider the C&LEX(X, Y, C) constraint where the C is REGULAR($\mathcal{A}, X$) and $\mathcal{A}$ is the automaton presented in Figure 1. Domains of variables are $X[1] \in \{1, 2\}$, $X[2] \in \{1, 3\}$, $X[3] \in \{2\}$ and $Y[1] \in \{1, 2, 3\}$, $Y[2] \in \{1, 2\}, Y[3] \in \{1, 3\}$.

Consider Algorithm 1 that finds the lexicographically smallest solution of the REGULAR constraint. At line 7 it invokes a DC propagator for the REGULAR constraint to ensure that an extension of a partial solution on each step leads to a solution of the constraint. To do so, it prunes all values that are inconsistent with the current partial assignment. We will show that for the REGULAR constraint values consistent with the current partial assignment can be found in $O(log(d))$ time.

Let G_x be a layered graph for the REGULAR constraint and $L_i = [L[1], \ldots, L[i]]$ be a partial assignment at the ith iteration of the loop (lines 4 - 6, Algorithm 6). Then L_i

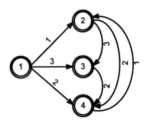

Fig. 1. Automation for Example 4

corresponds to a path from the initial node at 0th layer to a node q_j^i at ith layer. Clearly, values of $X[i+1]$ consistent with the partial assignment L_i are labels of outgoing arcs from the node q_j^i. We can find the label with the minimal value in $O(log(d))$ time. Algorithm 6 shows pseudo-code for $\text{REGULAR}_{min}(\mathcal{A}, L, X)$. Figure 2 shows a run of $\text{REGULAR}_{min}(\mathcal{A}, L, X)$ for variables X in Example 4. The lexicographically smallest solution corresponds to dashed arcs.

Algorithm 6. $\text{REGULAR}_{min}(\mathcal{A}, L, X)$

1: **procedure** $\text{REGULAR}_{min}(\mathcal{A} : in, L : out, X : in)$
2:     Build graph G_x;
3:     $q[0] = q_0^0$;
4:     **for** $i = 1$ **to** n **do**
5:         $L[i] = \min\{v_j | v_j \in outgoing\_arcs(q[i-1])\}$;
6:         $q[i] = t_{\mathcal{A}}(q[i-1], L[i])$; $\triangleright t_{\mathcal{A}}$ is the transition function of $\mathcal{A}$.
7:     **return** L;

The time complexity of Algorithm 2 for the REGULAR constraint is also $O(nT)$. The algorithm works with the layered graph rather than original variables. On each step it marks edges that occur in feasible paths in G_x that are lexicographically greater than or equal to L. Figure 3 shows execution of $\text{C\&LEX}_{lb}(X_l, Y, \text{REGULAR})$ for variables Y and the lexicographically smallest solution for X, $X_l = (1, 3, 2)$, from Example 4. It starts at initial node s and marks all arcs on feasible paths starting with values greater than $X_l[1] = 1$ (that is 2 or 3). Figure 3(a) shows the removed arc in gray and marked arcs in dashed style. Then, from the initial node at 0th layer it moves to the 2nd node at the 1st layer (Figure 3 (b)). The algorithm marks all arcs on paths starting with a prefix greater than $[X_l[1], X_l[2]] = [1, 3]$. There are no such feasible paths. So the $MarkConsistentArcs$ algorithm does not mark extra arcs. Finally, it finds that there is no outgoing arc from the 2nd node at 2nd layer labeled with 3 and stops its marking phase. There are two unmarked arcs that are solid gray arcs at Figure 3 (b). The algorithm prunes value 1 from the domain of $Y[1]$, because there are no marked arcs labeled with value 1 for $Y[1]$. Algorithm 8 shows the pseudo-code

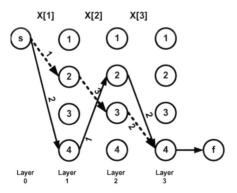

Fig. 2. The $\text{REGULAR}_{min}(\mathcal{A}, L, X)$ algorithm. Dashed arcs correspond to the lexicographically smallest solution.

Algorithm 7. Mark consistent arcs

1: **procedure** $MarkConsistentArcs(G_x : out, q : in)$
2:     Mark all arcs that occur on a path from q to the final node;

Algorithm 8. C&LEX$_{lb}(L, X, \text{REGULAR})$

1: **procedure** C&LEX$_{lb}(L : in, X : out, \text{REGULAR} : in)$
2:     Build graph G_x;
3:     $q[0] = q_0^0$;
4:     $q_L = 0$;
5:     **for** $i = 1$ **to** n **do**
6:         Remove outgoing arcs from the node $q[i-1]$ labeled with $\{min(X[i]), \dots, L[i]\}$;
7:         $MarkConsistentArcs(G_x, q[i-1])$;
8:         **if** ($\exists$ a outgoing arc from $q[i-1]) \wedge (i \neq 1)$ **then**
9:             mark arcs $(q[k-1], q[k]), k = q_L, \dots, i-1$;
10:            $q_L = i - 1$;
11:        **if** $L[i] \notin D(X[i])$ **then**
12:            break;
13:        $q[i] = T_{\mathcal{A}}(q[i-1], L[i])$;   ▷ $T_{\mathcal{A}}$ is the transition function of $\mathcal{A}$.
14:    **if** ($i == n$) **then**
15:        mark arcs $(q[k-1], q[k]), k = q_L, \dots, n$;
16:    **for** $i = 1$ **to** n **do**
17:        $Prune(\{v_j \in D(X[i]) | unmarked(v_j)\})$;

---

for C&LEX$_{lb}(L, X, \text{REGULAR})$. Note that the $MarkConsistentArcs$ algorithm for the REGULAR constraint is incremental. The algorithm performs a constant number of operations (deletion, marking) on each edge. Therefore, the total time complexity is $O(nT)$ at each invocation of the C&LEX$_{lb}(L, X, \text{REGULAR})$ constraint.

The second algorithm that we propose represents the C&LEX$(X, Y, \text{REGULAR})$ as a single automaton that is the product of automata for two REGULAR constraints and an automaton for LEX. First, we create individual automata for each of three constraints. Let Q be the number of states for each REGULAR constraint and d be the number of states for the LEX constraint. Second, we interleave the variables X and Y, to get the sequence $X[1], Y[1], X[2], Y[2], \dots, X[n], Y[n]$. The resulting automaton is a product of individual automata that works on the constructed sequence of interleaved variables.

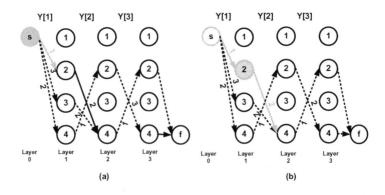

Fig. 3. A run of the C&LEX$_{lb}(L, X, \text{REGULAR})$ algorithm. Dashed arcs are marked.

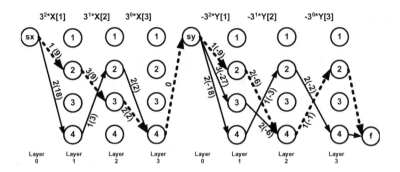

Fig. 4. The C&LEX(X, Y, REGULAR) algorithm. Dashed arcs show the shortest path through the arc $(sy, 2)$.

The number of states of the final automaton is $Q' = O(dQ^2)$. The total time complexity to enforce DC on the C&LEX(X, Y, REGULAR) constraint is thus $O(nT')$, where T' is the number of transitions of the product automaton. It should be noted that this algorithm is very easy to implement. Once the product automaton is constructed, we encode the REGULAR constraint for it as a set of ternary transition constraints [10].

The third way to propagate the C&LEX(X, Y, REGULAR) constraint is to encode it as a cost REGULAR constraint. W.L.O.G., we assume that there exist only one initial and one final state. Let G_x be the layered graph for REGULAR(X) and G_y be the layered graph for REGULAR(Y). We replace the final state at $n + 1$th layer in G_x with the initial state at 0th layer at G_y. Finally, we need to encode LEX(X, Y) using the layered graph. We recall that the LEX(X, Y) constraint can be encoded as an arithmetic constraint $(d^{n-1}X[1] + \ldots + d^0X[n] \le d^{n-1}Y[1] + \ldots + d^0Y[n])$ or $(d^{n-1}X[1] + \ldots + d^0X[n] - d^{n-1}Y[1] - \ldots - d^0Y[n] \le 0)$, where $d = |\bigcup_{i=1}^{n} D(X[i])|$.

In turn this arithmetic constraint can be encoded in the layered graph by adding weights on corresponding arcs. The construction for Example 4 is presented in Figure 4. Values in brackets are weights to encode the LEX(X, Y) constraint. For instance, the arc $(sx, 2)$ has weight 9. The arc corresponds to the first variable with the coefficient d^2, $d = 3$. It is labeled with value 1. The weight equals $1 \times d^2 = 9$. More generally, an arc between the $k - 1$th and kth layers labeled with v_j is given weight $v_j d^{n-k}$. Note that the weights of arcs that correspond to variables Y are negative. Hence, the C&LEX(X, Y, REGULAR) constraint can be encoded as a cost REGULAR($[\mathcal{A}, \mathcal{A}], [X, Y], W$) constraint, where W is the cost variable, $[\mathcal{A}, \mathcal{A}]$ are two consecutive automata. W has to be less than or equal to 0. Consider for example the shortest path through the arc $(sy, 2)$. The cost of the shortest path through this arc is 3. Consequently, value 1 can be pruned form the domain of $Y[1]$.

The time complexity of enforcing DC on the cost REGULAR($[\mathcal{A}, \mathcal{A}], [XY], W$) is $O(nT)$, where $d = |\bigcup_{i=1}^{n} D(X[i])|$ and T is the number of transitions of $\mathcal{A}$.[1] Again, the use of large integers adds a linear factor to the complexity, so we get $O(n^2T)$.

---

[1] Note that we have negative weights on arcs. However, we can add a constant d^n to the weight of each arc and increase the upper bound of W by this constant.

4   Experimental Results

To evaluate the performance of the proposed algorithms we carried out a series of experiments nurse scheduling problems (NSP) for C&LEX(X, Y, SEQUENCE) and C&LEX(X, Y, REGULAR) constraints. We used Ilog 6.2 for our experiments and ran them on an Intel(R) Xeon(R) E5405 2.0Ghz with 4Gb of RAM. All benchmarks are modeled using a matrix model of $n \times m$ variables, where m is the number of columns and n is the number of rows.

The C&LEX(X, Y, SEQUENCE) **constraint.** The instances for this problem are taken from www.projectmanagement.ugent.be/nsp.php. For each day in the scheduling period, a nurse is assigned to a day, evening, or night shift or takes a day off. The original benchmarks specify minimal required staff allocation for each shift and individual preferences for each nurse. We ignore these preferences and replace them with a set of constraints that model common workload restrictions for all nurses. Therefore we use only labor demand requirements from the original benchmarks. We also convert these problems to Boolean problems by ignoring different shifts and only distinguishing whether the nurse does or does not work on the given day. The labor demand for each day is the sum of labor demands for all shifts during this day. In addition to the labor demand we post a single SEQUENCE constraint for each row. We use a static variable ordering that assigns all columns in turn starting from the last one. Each column is assigned from the bottom to the top. This tests if propagation can overcome a poor branching heuristic which conflicts with symmetry breaking constraints. We used six models with different SEQUENCE constraints posed on rows of the matrix. Each model was run on 100 instances over a 28-day scheduling period with 30 nurses. Results are presented in Table 1. We compare C&LEX(X, Y, SEQUENCE) with the decomposition into two SEQUENCE constraints and LEX. In the case of the decomposition we used two algorithms to propagate the SEQUENCE constraint. The first is the decomposition of the SEQUENCE constraint into individual AMONG constraints (AD), the second is the original $HPRS$ filtering algorithm for SEQUENCE[2]. The decompositions are faster on easy instances that have a small number of backtracks, while they can not solve harder instances within the time limit. Overall, the model with the C&LEX(X, Y, SEQUENCE) constraint performs about 4 times fewer backtracks and solves about 80 more instances compared to the decompositions.

The C&LEX(X, Y, REGULAR) **constraint.** We implemented the second algorithm from Section 3.3, which propagates C&LEX(X, Y, REGULAR) using a product of automata for two REGULAR constraints and the automaton for the LEX constraint. C&LEX(X, Y, REGULAR) was compared with decomposition into individual REGULAR and LEX constraints. We used two models with different REGULAR constraints posed on rows of the matrix. Each model was run on 100 instances over a 7-day scheduling period with 25 nurses. We use the same variable ordering as above. The REGULAR constraint in the first model expresses that each nurse should have at least 12 hours of break between 2 shifts. The REGULAR constraint in the second model expresses that each nurse should have at least 12 hours of break between 2 shifts and at least

---

[2] We would like to thank Willem-Jan van Hoeve for providing us with the implementation of the $HPRS$ algorithm.

Table 1. Simplified NSPs. Number of instances solved in 60 sec / average time to solve.

	AD, Lex	$HPRS$, Lex	C&Lex
1 Sequence(3,4,5)	46 / 1.27	46 / 2.76	**74 / 1.44**
2 Sequence(2,3,4)	66 / 0.63	66 / 1.29	**83 / 2.66**
3 Sequence(1,2,3)	20 / 0.54	20 / 1.04	**34 / 3.17**
4 Sequence(4,5,7)	78 / 1.36	77 / 2.31	**82 / 2.43**
5 Sequence(3,4,7)	55 / 0.55	55 / 1.07	**58 / 1.53**
6 Sequence(2,3,5)	19 / 5.38	18 / 8.27	**31 / 1.74**
solved/total	284 /600	282 /600	**362 /600**
avg time for solved	1.230	2.194	**2.147**
avg bt for solved	18732	16048	**4382**

Table 2. NSPs. Number of instances solved in 60 sec / average time to solve.

	Regular, Lex	C&Lex
12 hours break	30 / 9.31	**93 / 2.59**
12 hours break + 2 consecutive shifts	87 / 1.05	**88 / 0.22**
solved/total	117 /200	**181 /200**
avg time for solved	3.166	**1.439**
avg bt for solved	35434	**1220**

two consecutive days on any shift. Results are presented in Table 2. The model with the C&Lex(X, Y, Regular) constraint solves 64 more instances than decompositions and shows better run times and takes fewer backtracks.

5   Related and Future Work

Symmetry breaking constraints have on the whole been considered separately to problem constraints. The only exception to this of which we are aware is a combination of lexicographical ordering and sum constraints [11]. This demonstrated 'that on more difficult problems, or when the branching heuristic conflicted with the symmetry breaking, the extra pruning provided by the interaction of problem and symmetry breaking constraints is worthwhile. Our work supports these results. Experimental results show that using a combination of Lex and other global constraints achieves significant improvement in the number of backtracks and run time. Our future work is to construct a filtering algorithm for the conjunction of the Hamming distance constraint with other global constraints. This is useful for modeling scheduling problems where we would like to provide similar or different schedules for employees. We expect that performance improvement will be even greater than for the C&Lex constraint, because the Hamming distance constraint is much tighter than the Lex constraint.

References

1. Puget, J.F.: On the satisfiability of symmetrical constrained satisfaction problems. In: Komorowski, J., Raś, Z.W. (eds.) ISMIS 1993. LNCS, vol. 689, pp. 350–361. Springer, Heidelberg (1993)
2. Walsh, T.: Breaking value symmetry. In: Bessière, C. (ed.) CP 2007. LNCS, vol. 4741, pp. 880–887. Springer, Heidelberg (2007)
3. Flener, P., Frisch, A.M., Kzlltan, B.H.Z., Miguel, I., Walsh, T.: Matrix modelling: Exploiting common patterns in constraint programming. In: Proc. of the Int. Workshop on Reformulating Constraint Satisfaction Problems, pp. 27–41 (2002)
4. Flener, P., Frisch, A., Hnich, B., Kiziltan, Z., Miguel, I., Pearson, J., Walsh, T.: Breaking row and column symmetries in matrix models. In: Van Hentenryck, P. (ed.) CP 2002. LNCS, vol. 2470, pp. 462–476. Springer, Heidelberg (2002)
5. Frisch, A., Hnich, B., Kiziltan, Z., Miguel, I., Walsh, T.: Global constraints for lexicographic orderings. In: Van Hentenryck, P. (ed.) CP 2002. LNCS, vol. 2470, pp. 93–108. Springer, Heidelberg (2002)
6. Beldiceanu, N., Contejean, E.: Introducing global constraints in CHIP. Mathematical and Computer Modelling 12, 97–123 (1994)
7. Pesant, G.: A regular language membership constraint for finite sequences of variables. In: Wallace, M. (ed.) CP 2004. LNCS, vol. 3258, pp. 482–495. Springer, Heidelberg (2004)
8. Carlsson, M., Beldiceanu, N.: Arc-consistency for a chain of lexicographic ordering constraints. TR T–2002-18, Swedish Institute of Computer Science (2002)
9. van Hoeve, W.-J., Pesant, G., Rousseau, L.-M., Sabharwal, A.: Revisiting the sequence constraint. In: Benhamou, F. (ed.) CP 2006. LNCS, vol. 4204, pp. 620–634. Springer, Heidelberg (2006)
10. Quimper, C.-G., Walsh, T.: Global Grammar constraints. In: Benhamou, F. (ed.) CP 2006. LNCS, vol. 4204, pp. 751–755. Springer, Heidelberg (2006)
11. Hnich, B., Kiziltan, Z., Walsh, T.: Combining symmetry breaking with other constraints: lexicographic ordering with sums. In: Proc. of the 8th Int. Sym. on the Artificial Intelligence and Mathematics (2004)

Iterative Flattening Search on RCPSP/max Problems: Recent Developments

Angelo Oddi and Riccardo Rasconi

Institute of Cognitive Science and Technology, CNR, Rome, Italy
{angelo.oddi,riccardo.rasconi}@istc.cnr.it

Abstract. This paper proposes an iterative improvement algorithm for solving instances of the Resource Constraint Project Scheduling Problem with Time-Windows (RCPSP/max). The algorithm is based on Iterative Flattening Search (IFS), an effective meta-heuristic strategy proposed over the past years for solving multi-capacity optimization scheduling problems. Given an initial solution, IFS iteratively applies two steps: (1) a subset of solving decisions are randomly retracted from a current solution (*relaxation-step*); (2) a new solution is incrementally recomputed (*flattening-step*). At the end, the best solution found is returned. To the best of our knowledge this is the first paper which proposes a version of IFS for solving RCPSP/max instances. The main contribution of this paper is threefold: (1) we succeed in improving *15* out of *90* solutions with respect to the officially published current best, thus demonstrating the general efficacy of IFS; (2) we highlight an intrisic limitation of the original IFS strategy in solving RCPSP/max, such that under certain circumstances the two-step improvement loop can get stuck in a status where no solving decision can be retracted; (3) we propose two different *escaping strategies* which extend the original IFS procedure. An experimental evaluation ends the paper, comparing the performances of the proposed escaping strategies against the original IFS procedure.

1 Introduction

This paper explores the solving capabilities of the Iterative Flattening Search (IFS) algorithm against scheduling problem instances belonging to the class of Resource Constrained Project Scheduling Problem with Time Windows (RCPSP/max). IFS represents a family of stochastic local search ([1]) techniques that was originally introduced in [2] as a non-systematic approach to solve difficult scheduling problem instances; as demonstrated in [3], RCPSP/max problems indeed belong to this category, as both the optimization and the feasibility versions of the problem are *NP-hard*.

IFS is devised to iteratively use heuristics for solving makespan-minimization scheduling problems, and it has been shown to have very good scaling capabilities. The procedure basically iterates two solving steps: (1) a *relaxation step*, where a subset of solving decision made at iteration $(i - 1)$ are randomly retracted at iteration i, and (2) a *flattening step*, where a new solution is re-computed after the previous relaxation. The choice of the term "flattening" stems from the fact that finding a solution equates to pushing down the resource usage profiles below the maximum capacity threshold of each resource involved in the problem (see Section 3.3).

A. Oddi, F. Fages, and F. Rossi (Eds.): CSCLP 2008, LNAI 5655, pp. 99–115, 2009.

A later improvement to the IFS algorithm was proposed in [4], where the original single relaxation step was replaced by an iterative version, obtaining significant improvements in solution quality against comparable computational efficiency. Additional optimal solutions and improvements on known upper-bounds for a Multi-Capacity Job Shop Scheduling Problem (MCJSSP) benchmarks were obtained in [5]: this approach follows the IFS schema but uses different engines for the flattening and relaxation steps.

This paper presents the first attempt, to the best of our knowledge, to use the IFS algorithm to solve RCPSP/max problem instances. The present work aims at widening the knowledge about IFS performance capabilities, by describing IFS's behavior against a particular situation (later on referred to as *stall*) into which the procedure may fall during the solving process. The stall situation is characterized by the impossibility to carry on any further relaxation step due to the absence of possibly retractable decisions along the solution's critical path. The occurrence of stalls has been discovered and analyzed; in this work, two different relaxation policies (not based on the critical path) are proposed to counter the stall effect, and their performance is assessed with respect to the "no-action" policy of the original IFS procedure. In case of stall, such policies basically look "somewhere else" in the current solution for alternative solving decisions to retract, hence helping the *retract-flattening* cycle to escape the deadlock and proceed towards an improved solution. As the paper will show, searching elsewhere for decisions to retract yields better results with respect to the approach that ignores the stall; moreover, the trend exhibited by the empirical results convinces us that pushing in the direction of more informed retraction heuristics may pay off in terms of solution quality.

The paper is organized as follows: in Section 2, the reference RCPSP/max problem is presented, as well as the benchmark set chosen for the performed experiments; Section 3 is dedicated to a thorough description of the IFS solving policy used throughout this work; Section 4 describes in detail the stall situation, as well as the limits exhibited by the original IFS implementation in tackling the resulting deadlock. Two novel strategies to escape the stall are introduced, as well as the algorithm that integrates them in the original IFS procedure. In Section 5, the experiments are described, and the most interesting results are explained. Section 6 finally concludes the paper.

2 Reference Problem and Benchmarks

The Resource Constrained Project Scheduling Problem (RCPSP) has been widely studied in Operations Research (OR) literature (see [6] for a survey). The RCPSP version with *Time Windows* (RCPSP/max) is an extended formulation of the basic problem which underlies a number of scheduling applications [7] and is considered particularly difficult, due to the presence of temporal separation constraints (in particular maximum time lags) between project activities.

2.1 The RCPSP/max

The RCPSP/max can be formalized as follows:

- a set V of n activities must be executed, where each activity A_i has a fixed duration p_i. Each activity has a start-time $st(A_i)$ and a completion-time $et(A_i)$ that satisfies the constraint $st(A_i) + p_i = et(A_i)$.

- a set E of temporal constraints exists between various activity pairs $\langle A_i, A_j \rangle$ of the form $st(A_j) - st(A_i) \in [T_{ij}^{min}, T_{ij}^{max}]$, called start-to-start constraints (time lags or *generalized precedence relations* between activities).[1]
- a set R of renewable resources are available, where each resource r_k is characterized by a maximum integer capacity $c_k \geq 1$.
- execution of an activity A_i requires some capacity from one or more resources. For each resource r_k the integer $rc_{i,k}$ represents the required capacity (or *size*) of activity A_i.

A schedule S is an assignment of values to the start-times of all activities in V ($S = (st(A_1), \ldots, st(A_n))$. A schedule is *time-feasible* if all temporal constraints are satisfied (all constraints $st(A_j) - st(A_i) \in [T_{ij}^{min}, T_{ij}^{max}]$ and $st(A_i) + p_i = et(A_i)$ hold). A schedule is *resource-feasible* if all resource constraints are satisfied (let $A(S,t) = \{A_i \in V | st(A_i) \leq t < st(A_i) + p_i\}$ be the set of activities which are in progress at time t and $r_k(S,t) = \sum_{A_i \in A(S,t)} rc_{i,k}$ the usage of resource r_k at that same time; for each t the constraint $r_k(S,t) \leq c_k$ must hold). A schedule is *feasible* if both sets of constraints are satisfied. Solving the RCPSP/max optimization problem equates to finding a feasible schedule with *minimum makespan* Mk, where $Mk(S) = max_{A_i \in V} \{et(A_i)\}$.

The feasibility version of the RCPSP is polynomial, while the optimization version has been demonstrated to be NP-hard and among the most intractable combinatorial optimization problems (see [8]). As opposed to the RCPSP, both the feasibility and the optimization versions of the RCPSP/max are NP-hard (see [3]), which makes the RCPSP/max an extrememly complex scheduling problem.

2.2   The UBO-200 Benchmarks

The RCPSP/max benchmarks that have been chosen for the present investigation are taken from the well known UBO test sets[2], generated by the project generator Pro-Gen/max ([9]). The UBO problems represent a rather unexplored and challeging benchmark set, whose size generally ranges from 10 to 1000 activities; the otpimal solutions for many instances of this benchmark are still unknown. in particular, the experiments have been performed with the UBO-200 benchmark set, composed of 90 RCPSP/max instances each made up of 200 activities and 5 multicapacity resources. It should be noted that the UBO-200 problems represent a rather difficult testbed for scheduling algorithms, because of the high number of activities involved, the complexity of the minimum and maximum temporal contraints network, as well as the complexity of the resource constraints arising from the distribution of the multicapacity resources among the activities. All these circumstances contribute to making the search space huge, and hence the need to employ efficient stochastic search algorithms.

---

[1] Note that since activity durations are constant values, end-to-end, end-to-start, and start-to-end constraints between activities can all be represented in start-to-start form.

[2] Available via world-wide-web at www.wior.uni-karlsruhe.de/LS_Neumann/ Forschung/ProGenMax/rcpspmax.html

3   Iterative Flattening Search

In this section we introduce a general IFS procedure, as depicted in Figure 1. The algorithm basically alternates relaxation and flattening steps until a better solution is found or a maximal number of iterations with no makespan improvement is executed. The procedure takes two parameters as input: (1) an initial solution S; (2) a positive integer $MaxFail$ which specifies the maximum number of non-makespan improving moves that the algorithm will tolerate before terminating. After initialization (Steps 1-2), a solution is repeatedly modified within the while loop (Steps 3-10) by the application of the RELAX and FLATTEN procedures. In case a better makespan solution is found (Step 6), the new solution is stored in S_{best} and the *counter* is reset to 0. Otherwise, if no improvement is found within $MaxFail$ moves, the algorithm terminates and returns the best solution found.

A first distinctive aspect of the IFS algorithm is that it is based on a basic *constructive search*. In previous works like [2,10] such feature is implemented as a *Precedence Constraint Posting* (PCP) algorithm where a set of solution precedence constraints is increasingly created while reasoning on resource contention peaks (see Section 3.3). Both works select precedences by means of a basic Earliest Start Time Algorithm (ESTA).

The second distinctive feature of the IFS is represented by the previously mentioned *relaxation step*, whose role is to relax a feasible solution into a possibly resource infeasible, but precedence feasible, schedule by removing some search decisions represented as precedence constraints between pair of activities; this schema integrates naturally with the formalism used in this work to represent a solution, that is the solution representation used by a PCP greedy algorithm.

The current section proceeds as follows: Section 3.1 is dedicated to the formalism employed to represent both the scheduling solutions and the temporal problem underlying every problem instance; Section 3.2 will describe the IFS relaxation step in details, while the flattening step will be the object of Section 3.3.

IFS($S,MaxFail$)
begin
1. $S_{best} \leftarrow S$
2. $counter \leftarrow 0$
3.   **while** ($counter \leq MaxFail$) **do**
4.   RELAX(S)
5. $S \leftarrow$ FLATTEN(S)
6.   **if** Mk(S) < Mk(S_{best}) **then**
7. $S_{best} \leftarrow S$
8.   counter $\leftarrow$ 0
9.   **else**
10.   counter $\leftarrow$ counter + 1
11. **return** (S_{best})
end

Fig. 1. The IFS general schema

3.1 The Scheduling Problem Representation Formalism

The class of scheduling algorithms we are focusing upon in this paper is based on a representation of the basic scheduling problem as a precedence graph $G(A, E)$ where A is the set of activities (plus two fictitious activities source a_{source} and sink a_{sink}), and E is the set of precedence constraints defined among the nodes in A. A solution S is represented as an extended graph G_S of G, characterized by an additional set of precedence constraints (or *decisions*) that are necessary to "solve" the original problem. More specifically, the set E is partitioned in two subsets, $E = E_{prob} \cup E_{post}$, where E_{prob} is the set of precedence constraints originating from the problem definition, and E_{post} is the set of precedence constraints posted to resolve resource conflicts. It should be noted that in general, the directed graph $G_S(A, E)$ does not represent a single solution but rather a *set* of temporal solutions.

Temporal reasoning procedures on each scheduling solution are based on a directed graph $TM(TP, E)$ called *time map* [11], where the set of nodes TP represents time-points or temporal variables (i.e., the origin point, the horizon point and the start and end time points, $st(A_i)$ and $et(A_i)$, of each activity A_i, and the set of edges E represents temporal distance constraints between pairs of time-points. Every temporal constraint has the general form $a \le tp_j - tp_i \le b$ and is represented in the graph $TM(TP, E)$ as a direct edge (tp_i, tp_j) with label $[a, b]$. Each time point $tp_i \in TP$ is associated to an interval $[lb_i, ub_i]$ of the possible time instants, or temporal values, where the event associated to the time-points may take place in time. The time-point tp_0, the origin point, is associated to the constant interval $[0, 0]$. The graph $TM(TP, E)$ corresponds to a *Simple Temporal Problem* (STP) [12]; the computation of the intervals $[lb_i, ub_i]$ and the check for the STP's consistency (an STP problem is inconsistent when there exists at least an empty interval $[lb_i, ub_i]$), can be polynomially determined via *shortest path* computations on a directed graph $G_d(V_d, E_d)$ called *distance graph*.

The graph G_d is obtained from the time map $TM(TP, E)$ as follows: (a) the set of nodes $V_d = TP$; (b) the set of edges E_d is built from the set E considering for each constraint $a \le tp_j - tp_i \le b \in E$ two weighted edges in the set E_d: the first one directed from tp_i to tp_j with weight b, the second one directed from tp_j to tp_i with weight $-a$. In G_d, the usual definitions of path and path's length on a weighted graph are assumed: a *path* is a sequence of consecutive edges $(tp_1, tp_2), (tp_2, tp_3) \ldots (tp_{n-1}, tp_n)$; the *length* of a path is the sum of the weights associated to the sequence of edges. A negative cycle is a closed path with negative length and an STP is consistent iff there are no negative cycles in its graph G_d [12]. When no negative cycle is contained in the graph, for each pair of time points (tp_i, tp_j), a shortest path distance d_{ij} is defined, and the constraint $-d_{ji} \le tp_j - tp_i \le d_{ij}$ holds. In particular, given that d_{i0} is the length of the shortest path on G_d from the time point tp_i to the origin point tp_0 and d_{0i} is the length of the shortest path from the origin point tp_0 to the time point tp_i, the interval $[lb_i, ub_i]$ of time values associated to the generic time variable tp_i is computed on the graph G_d as the interval $[-d_{i0}, d_{0i}]$ (see [12]).

3.2 The Relaxation Procedure

In general, a relaxation procedure transforms a feasible schedule into a possibly resource infeasible, but temporally feasible one, by adopting different strategies for

PCRELAX$(S, p_r, MaxRlxs)$
begin
1. **for** 1 **to** $MaxRlxs$
2.   **forall** $(et(A_i), st(A_j)) \in$ CriticalPath$(S) \cap E_{post}$
3.   **if** random$(0,1) < p_r$
4.   **then** $S \leftarrow S \setminus (et(A_i), st(A_j))$
end

Fig. 2. pc-based relaxation procedure

removing some search decisions. The strategy presented in this section, used in [2,4], removes precedence constraints[3] between pair of activities on the critical path (see below) of a solution. The computation of the critical path is required every time *before* launching the relaxation procedure. The reader should also note that in case of consecutive relaxations, retracting a set of constraints at iteration $(i - 1)$ generally implies a total modification of the previous critical path, which therefore has to be re-computed from scracth, before retracting the next constraint set at iteration i. In this section we consider the iterated relaxing strategy as presented in [4].

As mentioned above, the relaxation step is based on the concept of *critical path*. Resuming the definition of path and of path length given in Section 3.1, given a scheduling solution S, the critical path of S is a sequence of consecutive edges $(tp_1, tp_2), (tp_2, tp_3)$ ... (tp_{m-1}, tp_m) where $tp_1 = et(a_{source})$ and $tp_m = st(a_{sink})$, such that any increase in the length of the critical path directly reflects in an equivalent increase in S's makespan. Therefore, any improvement in makespan will necessarily require change to some subset of precedence constraints situated on the *critical path*, since these constraints collectively determine the solution's current makespan. Following this observation, the relaxation step introduced in [2] is designed to retract some number of precedence constraints posted on the solution's critical path. Figure 2 shows the PCRELAX procedure. Steps 2-4 consider the set of posted precedence constraints $(et(A_i), st(A_j))$, which belong to the current critical path and the set E_{post}. A subset of these constraints is randomly selected on the basis of the parameter $p_r \in (0, 1)$ and then removed from the current solution.

3.3   The Flattening Procedure

The relaxation schema yields an intermediate solution containing resource contention peaks that should be *flattened* (removed from the current solution). To this aim, we have implemented a general solution schema, based on a Precedence Constraint Posting (PCP) strategy. The idea is the following: the existence of any resource contention peak is motivated by the overlapping in time of two or more activities that require the same resource, exceeding the resource's maximum capacity. It is straightforward that in order to solve the conflict, such activities should be temporally *separated*; the PCP strategy is based on posting those new constraints that are necessary to separate the activities

---

[3] Note that only precedence constraints belonging to the E_{post} set are retracted. Constraint relaxation *never* involves the original problem constraints.

```
PCPS(P, S)
begin
1.   Propagate(S)
2.   if IsSolution(S)
3.      then return(S)
4.      else
5.         mcs ← SelectConflict(P, S)
6.         if Solvable(mcs, S)
7.            then
8.               pc ← ChoosePrecedence(S, mcs)
9.               PCPS(P, S ∪ {pc})
10.           else return(fail)
end
```

Fig. 3. The PCPS algorithm

involved in any resource conflict, until all conflicts are eliminated and a temporally and resource feasible solution is found.

The flattening step (see Figure 3) used in [2] is inspired by prior work on the Earliest Start Time Algorithm (ESTA) from [13]. The algorithm is a variant of a class of PCP scheduling procedures characterized by a two-phase solution generation process. The first step *constructs an infinite capacity solution*. The current problem is formulated as an STP [12] temporal constraint network (see Section 3.1), where temporal constraints are modeled and satisfied (via constraint propagation), yielding a time feasible solution that assumes infinite resource capacity.

The second step *levels resource demand by posting precedence constraints*. Resource constraints are super-imposed by projecting "resource demand profiles" over time. The detected resource conflicts, once reduced to Minimal Conflict Sets (MCS), are then re-solved by iteratively posting simple precedence constraints between pairs of competing activities. A MCS (see [10,14] for further details), is defined as a *set of activities that simultaneously require a resource r_k with a combined capacity requirement $> c_k$, such that the combined requirement of any proper subset is $\leq c_k$*. From the previous definition it follows that any precedence constraint separating any two activities belonging to an MCS eliminates the resource conflict, therefore, isolating all the MCSs from a contention peak represents a great advantage. The constraint posting process of ESTA is based on the Earliest Start Solution (ESS) consistent with currently imposed temporal constraints. At Step 1 the procedure *Propagate* propagates the current temporal constraints. It then proceeds to compute a resource conflict (Steps 2-5). If this set is empty the ESS is also resource feasible and a solution is found; otherwise if a conflict exists that can be solved, a new precedence constraint is posted (Steps 8-9); otherwise the process fails (Step 10), meaning that the temporal constraints currently posted in the solution disallow the separation of any pair of activities belonging to the MCS (*unresolvable MCS*). For further details on the functions *SelectConflict()*, and *ChoosePrecedence()* (non deterministic version of the precedence selection operator) the reader should refer to the original references.

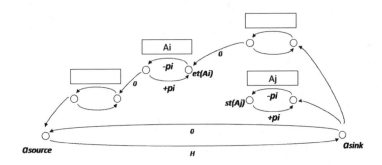

Fig. 4. The shortest path distance d_{ij} from $et(A_i)$ to $st(A_j)$ for RCPSPs

3.4   RCPSP/max: Remarks on Temporal Propagation Complexity

Turning from RCPSP to RCPSP/max introduces an overhead of complexity the reader should be made aware of, and which can help to explain the experimental results presented in Section 5. As explained above, within the IFS algorithm the calculus of the shortest path distances d_{ij} on the graph G_d plays a fundamental role in both the flattening step (when precedence constraints are added) and the relaxation step (when one or more precedence constraints are retracted). In particular, in the procedure shown in Figure 3, a precedence constraint is selected and posted between a pair of activities A_i and A_j when considered the general inequality $-d_{ji} \leq st(A_j) - et(A_i) \leq d_{ij}$ (see Section 3.1) the condition $d_{ij} \geq 0$ holds.

In the case of RCPSP instances, given a pair of *unordered* activities A_i and A_j, the distance d_{ij} between the two time points $et(A_i)$ and $st(A_j)$ can be calculated as $d_{ij} = d_{i0} + d_{0j}$, where d_{i0} is the shortest path distance from the time point $et(A_i)$ to the *source* and d_{0j} is the shortest path distance from the *source* to $st(A_j)$. The last statement can be proved with the help of Figure 4. In fact, given a generic RCPSP instance, the correspondent graph G_d contains three different kind of temporal constrains: the *horizon* constraint $0 \leq st(a_{sink}) - et(a_{source}) \leq H$; the duration constraints $p_i \leq et(A_i) - st(A_i) \leq p_i$ imposed on all the activities; the constraints posted between the activities, which have the form $a \leq st(A_j) - et(A_i) \leq +\infty$. As it is possible to verify on Figure 4, the only way to find a shortest path from $et(A_i)$ to $st(A_j)$ is to move backward towards the *source* (reference node with id 0), to follow the only maximal constraints H, and reach the time point $st(A_j)$ from the *sink* (the node with id $n + 1$), hence the length of the shortest path from $et(A_i)$ to $st(A_j)$ has the value $d_{ij} = d_{i0} + H + d_{(n+1)j} = d_{i0} + d_{0j}$. The shortest path information respect to the source time point can be maintained by means of a *Single Source Shortest Path* propagation algorithms (e.g. Bellman-Ford [15]), whose complexity is $O(|V| \cdot |E|)$, where $|V|$ is the number of time points and $|E|$ is the number of edges in the STP.

As opposed with RCPSP, in the RCPSP/max case, information about the distance between any pair of time points requires a more complex computation, because the shortest path between a generic pair of time points $st(A_j)$ and $et(A_i)$ is not constrained to contain the origin tp_0 as in the previous case. Therefore, such information is generally maintained as a $|V| \times |V|$ matrix called *distance matrix*, whose consistency maintenance

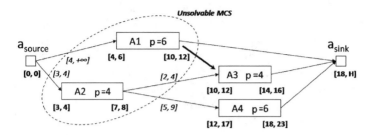

Fig. 5. An example of deadlock (stall)

requires the utilization of *All Pair Shortest Path* algorithms (e.g. Floyd-Warshall [15]), with an increase in computational complexity.

The second source of complexity regards the relaxation step, as in the RCPSP/max case, the need to keep the distance information among all the time points consistent and available after every constraint retraction, requires that the distance matrix be recomputed from scratch (e.g., by employing the Floyd- Warshall algorithm), an operation which has complexity $O(|V|^3)$, and that is critical, given the high number of relaxations that characterize the IFS algorithm.

4 Improving the Current IFS Algorithm

Iterative Flattening Search was demonstrated to be an effective and scalable method for scheduling optimization [2,4]. Our current research goal is to extend the simple and effective search paradigm proposed by IFS to different and more general scheduling problems than the original Resource Constrained Project Scheduling Problem (RCPSP). As introduced above, in this work we propose our analysis on RCPSP/max instances, where the presence of time-windows constraints makes even the search version of a resource feasible solution an *NP-hard* problem.

As it will be described in the following section about the experimental analysis, the original IFS algorithm [2,4] generally shows very interesting performance (see Figure 1) on RCPSP/max instances. Yet, while running the experiments we discovered that the original version of IFS also exhibits an intrinsic limitation.

The small example shown in Figure 5 clarifies the problem. The graph shows a single-resource scheduling problem, where all the activities (represented as boxes) require the same amount of resource which is equal to the maximum resource capacity; therefore, each pair of activities represents a Minimal Critical Set (MCS). In addition, the names of activities and their durations are represented inside the boxes (e.g., A_1 $p = 6$ meaning that the activity A_1 has duration equal to 6 time units), the temporal constraints imposed between the *end-time* and the *start-time* of two activities is represented as a labelled directed edge with label $[lb, ub]$ (*italic* fonts). When there is no label, the default value is $[0, +\infty]$; the interval of the possible start-times (end-times) of each activity is represented as an interval of bold-face values $[lb, ub]$ posted at the lower left-corner (lower right-corner) of each activity. Finally, two additional activities are depicted, the a_{source} representing the time origin, and the a_{sink}, representing the scheduling horizon; for these two activities we only show the bounds associated to the end-time and the start-time, respectively. In the example of Figure 5, let us suppose that the precedence constraint between the activites A_1 and A_3 has been posted in order to find a solution; the time bounds shown in the figure are updated as a consequence of the insertion of such precedence constraint. In particular, it is easy to see that the current makespan has value 18 time units, which corresponds to the *critical-path* represented by the sequence $(et(a_{source}), st(A_2)), (et(A_2), st(A_4)), (et(A_4), st(a_{sink}))$. After the insertion of the precedence constraint between A_1 and A_3, Figure 5 gives a snapshot of a particular situation that the IFS procedure may easily encounter during the solving process, characterized by the following conditions:

1. there are no solution precedence constraints on the current critical path, as can be easily confirmed by visual inspection;
2. the graph does not represent a solution, because there exists an unsolvable Minimal Critical Set (MCS) composed of the activities A_1 and A_2. This MCS is unsolvable because activity A_2 cannot start after than $t = 4$ and cannot end before $t = 7$, so the activity is forced to be executed in the interval $[4, 7)$; at the same time, A_1 cannot start after than $t = 6$ and cannot end before $t = 10$, so the activity is forced to be executed in the interval $[6, 10)$. As a consequence, the activities are forced to overlap in the interval $[6, 7]$ and they can by no means be separated.

This situation characterizes a stall (or deadlock) because condition 1 makes it impossible for the IFS procedure to remove any constraint, in the current as well as in all the possibly remaining relaxation steps. As a consequence of the impossibility to retract previous decisions, all the remaining flattening steps will be confronted with the same partial solution, which greatly increases the probability that the MCS is going to remain unsolved until the $MaxFail$ number of cycles will be exhausted.

There is a subtle point behind the previous issue, that should be better clarified. The FLATTEN step in the IFS algorithm (see Figure 1) can be implemented according to two different policies. In the first, the temporal constraints imposed during the flattening step are committed in the current partial solution *even when it does not represent a feasible solution* (i.e., the flattening step has failed); in the second policy, the temporal constraints are committed *only if the flattening step finds a feasible solution*. The issue is subtle because if the flattening step is implemented according to the first policy, the stall represents an unresolvable deadlock; looking at Figure 5, it is easy to see that in case the added precedence constraint between A_1 and A_3 is left in the partial solution, the resulting MCS will *never* be resolved. In the opposite case, since the precedence constraint is removed when no feasible solution is found, the flattening step of the IFS procedure retains some chances to spontaneously exit the stall, for instance by imposing a precedence constraint between a_{source} and A_1 that pushes the latter activity beyond A_3 and A_4.

Nevertheless, even in case the flattening step is implemented according to the second policy, the stall is not desirable for at least two reasons:

– as a matter of fact, the stall severely impairs the solving capabilities of the IFS algorithm in that it inhibits one of the two pillars the procedure rests upon: the

IFS(S,$MaxFail$)
begin
1. $S_{best} \leftarrow S$
2. $counter \leftarrow 0$
3.   **while** $(counter \leq MaxFail)$ **do**
4. $C \leftarrow$ RELAX(S)
5.   **if**($C = \emptyset$ **and** Unsolved(S))
6.   **then** Apply-Escaping-Strategy(S)
7. $S \leftarrow$ FLATTEN(S)
8.   **if** Mk(S) < Mk(S_{best}) **then**
9. $S_{best} \leftarrow S$
10.   counter $\leftarrow 0$
11.   **else**
12.   counter $\leftarrow$ counter + 1
13. **return** (S_{best})
end

Fig. 6. IFS search with *escaping strategy*

relaxation step. The stall forces the IFS procedure to merely "go forward", which completely spoils the efficiency of the algorithm;
– stalls are mostly encountered after a considerable number of solving cycles, when the makespan has already been significantly reduced. This considerably lowers the possibility that the flattening step will find an alternative solution (as described in [16]). Empirical evidence confirms in fact that once a stall is ecountered, the procedure hardly ever succeeds in escaping the deadlock.

For all these reasons, it is necessary to introduce a mechanism that, going beyond the original relaxation policy based on the critical path, may help the procedure to escape the deadlock and allow it to further improve the current best solution. Figure 6 shows an extended version of Iterative Flattening Search, which uses an additional *stall escape strategy* for managing cases like the one shown in Figure 5. The procedure mainly works like the original one presented in Figure 1; the main difference is located at Step 5 where, in case both the previous stall conditions are satisfied, an escape strategy is applied. In order to escape stalls, we propose two different strategies:

- *Full Restart* (IFS-FR): according to this strategy, when a stall is encountered the solution is immediately brought back to its original state. In other words, *all* the solution constraints imposed in previous FLATTEN steps are retracted, which corresponds exactly to starting the resolution from scratch;
- MCS-*based Restart* (IFS-MCSR): as opposed to the previous case, the rationale behind this strategy is that the solving efforts made so far should not be completely discarded. As a stall is encountered, the solving process investigates *alternative* paths for relaxable constraints. These new paths are computed on the basis of the unsolved MCS reported along the last failure of the FLATTEN algorithm. More specifically, for each pair of activities (A_i, A_j) belonging to an unsolvable MCS $= \{A_1, A_2, \ldots, A_m\}$, the shortest path sequences between $et(A_i)$

GETCRITICALDECISIONS(MCS)
begin
1. $L \leftarrow \emptyset$
2. **foreach** $(A_i, A_j) \in \{(A_i, A_j)|A_i \neq A_j \wedge A_i, A_j \in \text{MCS}\}$
3. $tp = st(A_j)$
4.   **do**
5. $tp_s = pred(et(A_i), tp)$
6.   **if** $Decision(tp_s, tp)$
7.   **then** $Push((tp_s, tp), L)$
8. $tp \leftarrow tp_s$
9.   **while** $tp_s \neq et(A_i)$
10. **return**(L)
end

Fig. 7. IFS-MCSR: computation of the critical decisions

and $st(A_j)$ are computed, and all the solution precedence constraints possibly contained in all such sequences are removed. Removing the constraints that combine to make the MCS unresolvable, increases the possibility that the MCS be "unlocked" and possibly resolved in future flattening steps.

Figure 7 shows the algorithm *getCriticalDecisions*, which considers as input an unsolvable MCS and returns the list L of decisions to remove. In particular, for each pair of activities (A_i, A_j) belonging to the unsolvable MCS, the procedure collects all the precedence constraints imposed along the shortest path from the time point $et(A_i)$ to $st(A_j)$ (the inner loop at Steps 4-9). The procedure works on a G_d representation of the temporal information where the shortest paths are represented via a predecessor function $pred()$, such that, $tp_z = pred(tp_x, tp_y)$ is the predecessor time point of tp_y along the shortest path from tp_x to tp_y, so the path takes the form $(tp_x, tp_1), (tp_2, tp_3) \ldots (tp_z, tp_y)$. In case a precedence constraint $0 \leq tp_s - tp \leq +\infty$ has been posted between the current (tp_s, tp) pair of time points ($Decision(tp_s, tp)$ is true), the pair (tp_s, tp) is collected in a list L.

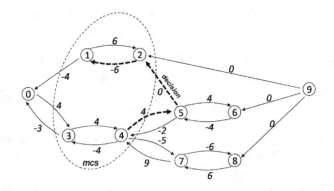

Fig. 8. G_d graph for the example of Figure 5

Figure 8 depicts the graph G_d related to the stall example already presented in Figure 5. As described above, the graph G_d is obtained from the original time map by trasforming each constraint $a \le tp_j - tp_i \le b \in E$ in two weighted edges: the first one from tp_i to tp_j with weight b (the edge is omitted when $b = +\infty$), the second one from tp_j to tp_i with weight $-a$. In particular, when we apply the procedure *getCriticalDecisions()* to the only unsolvable MCS $= \{A_1, A_2\}$, along the shotest path $(tp_4, tp_5), (tp_5, tp_2), (tp_2, tp_1)$ from $et(A_2) = tp_4$ to the $st(A_1) = tp_1$, the only decision collected is the one posted between tp_2 and tp_5.

The procedure depicted in Figure 6 represents a specific contribution of this paper, and in the following experimental section the original IFS strategy will be compared against the two new stall escape strategies.

5   Experimental Analysis

In this section we present the results of the set of experiments that have been performed in order to assess the behavior of the devised stall-escape strategies. The empirical analysis has been organized as follows: the whole set of RCPSP/max UBO-200 benchmark (90 instances) has been solved with the IFS algorithm shown in Figure 6. The $MaxRlxs$ parameter and the removal probability p_r of the relaxation step (see Figure 2), have been respectitely set to 4 and 0.2. The $MaxFail$ parameter of the algorithm (see Figure 1) has been set to 100; the choice of such a small value is motivated by the fact that the main objective of the experiments was to evaluate the capabilities of the algorithm to escape the stall situation, rather than pushing the IFS algorithm towards the improvement of the published makespan optima. Secondly, the authors were interested in measuring the convergence rate of the algorithm (i.e., the speed with which it converges towards the best known results), rather than its absolute performances. The reader should bear in mind that this work represents the first attempt to use the IFS procedure to solve large RCPSP/max instances against the possible occurrence of the stall situations; experiments with higher values of the $MaxFail$ parameter are currently being devised. All the experiments have been performed using the libraries provided by the Timeline-based Representation Framework (TRF, see [17]), a general modeling framework for Planning & Scheduling problem fast prototyping.

The efficacy of such strategies is compared against the best known results published at www.wior.uni-karlsruhe.de/LS_Neumann/Forschung/ProGenMax/rcpsp-max.html, and the original IFS behavior, which does not take the stall situation into account at all. The main results of the experiments are shown in Table 1.

Table 1. Summary of the main experimental results

	IFS	IFS-FR	IFS-MCSR
No. improved MKs	12	15	14
Avg. Makespan Gap	2.06	1.81	1.65
Avg. CPU time	2148.7	2024.7	1716.7
Avg. IFS cycles	142.7	148.5	145.0

In the table, the behavior of the three aforementioned strategies is compared according to the following metrics:

- *Number of Improved Makespans*: this figure represents the number of instances where the IFS algorithm succeeded in improving the makespan value w.r.t. the current best;
- *Average Makespan Gap* $[\Delta_{mk}^{avg}]$[4]: this metric returns the average makespan gap between the best published results (mk_i^0) and our experimental results, for all the n instances belonging to the UBO-200 set;
- *Average CPU Time*: this figure returns the average CPU Time employed by the IFS algorithm to solve all instances;
- *Average* IFS *Cycles*: this figure returns the average number of $\{relaxation - flatten\}$ cycles performed by IFS to solve all instances;

The first result that catches the eye is the number of makespan inprovements that has been obtained; on a total of 90 problems, the original version of the IFS algorithm succeeds in improving *12* instances (i.e., more than *13%*). This circumstance is remarkable, and a first conclusion can be drawn: the IFS approach can be effectively used also against the RCPSP/max benchmark; its efficacy is mainly proved by the fact that all experimental runs have been performed with an extremely low value of the $MaxFail$ parameter (*100*), where in general a value around the tenths of thousand is employed. Indeed, the presence of maximum temporal constraints in the problems does not seem to affect the algorithm's convergence speed.

As the table shows, the average time to obtain a solution when no stall exit strategy is employed is about *2150* seconds (around *36* minutes); the reader should not be misled by the high solving time. As explained in Section 3.3, tackling the RCPSP/max entails an increase in computational complexity, with respect to solving RCPSP instances. Nonetheless, a considerable boost might be obtained by using the fastest versions available of the All Pair Shortest Path propagation algorithms, which is outside the scope of this paper.

In this work, the attention should rather be focused on the low number of $\{relaxation - flatten\}$ cycles necessary to converge, because this figure is independent from all the previous factors, and therefore should be regarded as one of the fairest efficiency measure. As shown, the average number of solving cycles is between *140* and *150* for all strategies: as we will see, this is an interesting point, as it proves that using a different stall escape strategy does not affect (on average) the number of solving cycles, but does affect *how effectively* these cycles are employed. This last remark brings us directly to the next issue of our experimentation, namely, measuring any difference in performance that may depend on the different method used to counter the stalls. The table shows that using either the IFS-FR or the IFS-MCSR strategy increases number of improved instances, *15* (plus *16.6%*) and *14* (plus *15.6%*), respectively. Though the extent of such improvements may seem modest, this result clearly proves the existence of a

---

[4] Computed as:

$$\Delta_{mk}^{avg} = \frac{1}{n} \sum_{i=1}^{n} \frac{mk_i - mk_i^0}{mk_i^0}.$$

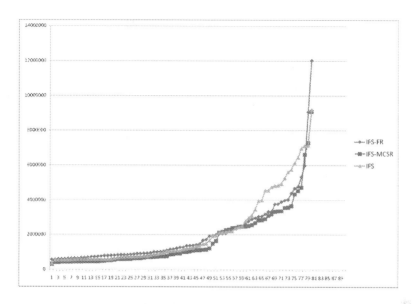

Fig. 9. CPU time spread accross the whole set of solved instances

quality increasing trend, and demonstrates that the utilization of stall escape heuristics does allow for a more thorough investigation of the search space.

Interestingly enough, such improvements are also obtained with a significant decrease of the average computing time. In fact, using the IFS-FR strategy is *5.7%* faster, while using the IFS-MCSR strategy is *20%* faster with respect to the original IFS procedure. Along with the previous results, it should be noticed that the average makespan gap (Δ_{mk}^{avg}) tends to decrease when stall escape strategies are employed: compared to the current best publicly available, such reduction is assessed to a value of *2.06, 1.81* and *1.65* when the IFS, IFS-FR and IFS-MCSR strategies are respectively used.

With this in mind, it is also possible to measure the performance of the IFS-FR and IFS-MCSR strategies against the original IFS policy. The figures obtained from this analysis (not shown on the table) demonstrate that the makespan values using IFS-FR are decreased on average by *-0.23*, and that the makespan values using the IFS-MCSR are decreased on average by *-0.38*, over the IFS strategy.

The results presented in Table 1 can be summarized as follows:

1. a significant number of makespan improvements have been found, which proves the general effectiveness of the IFS approach;
2. the utilization of stall escape policies yields better makespan optimization properties, either in terms of number and quality of improvements. More specifically, the IFS-MCSR strategy is more effective than the IFS-FR strategy;
3. the utilization of stall escape policies improves computational efficiency. More specifically, the IFS-MCSR strategy is more efficient than the IFS-FR strategy.

The last point is better described with the help of Figure 9, where the whole range of solved problem instances is represented in the *x*-axis, while the *y*-axis measures the solving CPU time. The solving time is depicted for all of the stall escape strategies employed, IFS-FR, IFS-MCSR and IFS. As the graph shows, for about half of the instances the three strategies barely employ the same time to find a solution (given that the IFS-MCSR always perform slightly better); but as the difficulty of the problem increases, the differences become more significant. In particular, in the second half of the x-axis it is evident that the utilization of a stall escape strategy pays off with respect to the policy that ignores the stalls. More in details, it can be seen that the IFS-MCSR generally exhibits a better performance with respect to the IFS-FR strategy.

6   Conclusions and Future Work

This work gives a first contribution about the role of IFS strategies in solving Resource Constraint Project Scheduling Problem with Time-Windows (RCPSP/max). RCPSP/max represents a hard and general scheduling problem, such that even the search of a feasible solution is *NP-hard*. In this paper we give two different contributions: firstly, we discover an intrinsic limitation of the original IFS strategy in solving RCPSP/max instances that may seriously affect IFS's performance in all cases where the typical IFS two-step improvement loop gets stuck in a state (stall) where no solving decision can be retracted; in order to exit the stall and as a second contribution, we propose two different *escaping strategies* which extend the original IFS procedure.

The performed experimental analysis has revealed that the originally proposed IFS strategy can be effective in solving large-size RCPSP/max instances and has paved the way for future research work on the definition of more effective solving procedures for large size benchmarks. As a further extension of the present work, we plan to study the effects of different flattening and relaxation procedures within the IFS loop, where key issues for improving effectiveness will be the use of efficient temporal and resource propagation algorithms as well as the definition of different strategies to retract decision constraints form the current solution.

Acknowledgments. Angelo Oddi and Riccardo Rasconi's work is partially supported by CNR under project RSTL (funds 2007) and ESA (European Space Agency) under the APSI initiative. The authors are grateful to Amedeo Cesta, Nicola Policella and Stephen F. Smith for their past joint work on Iterative Flattening Search and for their contribution to this work through interesting discussions and suggestions. The authors also wish to express their gratitude to all the members of the Planning and Scheduling Team (PST) in Rome, for providing a warm, friendship and stimulating environment which greatly helped the realization of this paper.

References

1. Hoos, H.H., Stützle, T.: Stochastic Local Search. In: Foundations and Applications. Morgan Kaufmann, San Francisco (2005)
2. Cesta, A., Oddi, A., Smith, S.F.: Iterative Flattening: A Scalable Method for Solving Multi-Capacity Scheduling Problems. In: AAAI/IAAI. 17th National Conference on Artificial Intelligence, pp. 742–747 (2000)

3. Bartusch, M., Mohring, R.H., Radermacher, F.J.: Scheduling Project Networks with Resource Constraints and Time Windows. Annals of Operations Research 16, 201–240 (1988)
4. Michel, L., Van Hentenryck, P.: Iterative Relaxations for Iterative Flattening in Cumulative Scheduling. In: ICAPS 2004. Proceedings of the 14th International Conference on Automated Planning & Scheduling, pp. 200–208 (2004)
5. Godard, D., Laborie, P., Nuitjen, W.: Randomized Large Neighborhood Search for Cumulative Scheduling. In: ICAPS 2005. Proceedings of the 15th International Conference on Automated Planning & Scheduling, pp. 81–89 (2005)
6. Brucker, P., Drexl, A., Mohring, R., Neumann, K., Pesch, E.: Resource-Constrained Project Scheduling: Notation, Classification, Models, and Methods. European Journal of Operational Research 112(1), 3–41 (1999)
7. Neumann, K., Schwindt, C.: Activity-on-Node Networks with Minimal and Maximal Time Lags and Their Application to Make-to-Order Production. Operation Research Spektrum 19, 205–217 (1997)
8. Schäffter, M.: Scheduling with respect to forbidden sets. Discrete Applied Mathematics 72, 141–154 (1997)
9. Kolisch, R., Schwindt, C., Sprecher, A.: Benchmark Instances for Project Scheduling Problems. In: Weglarz, J. (ed.) Handbook on Recent Advances in Project Scheduling. Kluwer, Dordrecht (1998)
10. Cesta, A., Oddi, A., Smith, S.F.: A constraint-based method for project scheduling with time windows. J. Heuristics 8(1), 109–136 (2002)
11. Dean, T.: Large-scale temporal data bases for planning in complex domains. In: IJCAI, pp. 860–866 (1987)
12. Dechter, R., Meiri, I., Pearl, J.: Temporal constraint networks. Artificial Intelligence 49, 61–95 (1991)
13. Cesta, A., Oddi, A., Smith, S.: Profile Based Algorithms to Solve Multiple Capacitated Metric Scheduling Problems. In: AIPS 1998. Proceedings of the 4th International Conference on Artificial Intelligence Planning Systems, pp. 214–223 (1998)
14. Laborie, P., Ghallab, M.: Planning with Sharable Resource Constraints. In: Proceedings of the 14th Int. Joint Conference on Artificial Intelligence (IJCAI 1995) (1995)
15. Cormen, T.H., Leiserson, C.E., Rivest, R.L., Stein, C.: Introduction to Algorithms, 2nd edn. MIT Press, Cambridge (2001)
16. Oddi, A., Cesta, A., Policella, N., Smith, S.F.: Iterative flattening search for resource constrained scheduling. J. Intelligent Manufacturing (2008) (published on-line November, 2008), doi:10.1007/s10845-008-0163-8
17. Cesta, A., Fratini, S.: The timeline representation framework as a planning and scheduling software development environment. (unpublished manuscript) (2008)

Robust Solutions in Unstable Optimization Problems

Maria Silvia Pini[1], Francesca Rossi[1], Kristen Brent Venable[1],
and Rina Dechter[2]

[1] Dipartimento di Matematica Pura ed Applicata, Università di Padova, Italy
{mpini,frossi,kvenable}@math.unipd.it
[2] School of Information and Computer Science, University of California,
Irvine, CA, USA
dechter@ics.uci.edu

Abstract. We consider constraint optimization problems where costs
(or preferences) are all given, but some are tagged as possibly unstable,
and provided with a range of alternative values. We also allow for some
uncontrollable variables, whose value cannot be decided by the agent
in charge of taking the decisions, but will be decided by Nature or by
some other agent. These two forms of uncertainty are often found in
many scheduling and planning scenarios. For such problems, we define
several notions of desirable solutions. Such notions take into account not
only the optimality of the solutions, but also their degree of robustness
(of the optimality status, or of the cost) w.r.t. the uncertainty present
in the problem. We provide an algorithm to find solutions accordingly
to the considered notions of optimality, and we study the properties of
these algorithms. For the uncontrollable variables, we propose to adopt
a variant of classical variable elimination, where we act pessimistically
rather than optimistically.

1 Introduction

Constraint programming [2,11] is successfully applied to many application domains. Constraint satisfaction problems are defined by decision variables, domains, and constraints that have to be satisfied. Optimization problems have an
objective function, or they associate costs, or preferences, with partial variable
instantiations, in order to discriminate among the possibly many solutions of the
problem.

Soft constraints [1] are a general formal modelling framework for constraint
optimization problems, where it is possible to express several different optimization criteria. For example, both fuzzy constraints and weighted constraints, as
well as MaxCSPs, can naturally be modelled in this formalism.

The specification of a complex constraint optimization problem is a difficult
modelling task, that tries to capture the current knowledge about the constraints
and the costs of the problem. Even when the specification is complete, only some
parts of the problem's parameters may be certain. Others may be viewed as
unstable due to possible future changes.

A. Oddi, F. Fages, and F. Rossi (Eds.): CSCLP 2008, LNAI 5655, pp. 116–131, 2009.
© Springer-Verlag Berlin Heidelberg 2009

Unstable costs are present in many real-life problems. A typical example is the budget estimate for next year in a company. Typically, such an estimate is based on data which is not known or not certain, and most of the times such uncertainty is represented by using last year's value (which can be seen as the default value), plus some range of possible other values around the default value. For example, one may not have the cost of fuel for next year, but he may know that last year it was 2 dollars, and usually the new value is never more than 30% higher. As another example, one may have to base some calculation on the number of pieces that will be produced in the year: a reasonable estimate could be last year's number assuming that the new number is within 5% from the old value. In the first example, the default value is at the lower end of the range, while in the second example it is in the middle of the range.

Other types of problems where unstable values may occur are when we want to numerically represent linguistic concepts, such as "more or less", "around", "at least", or "at most". In all these cases, the natural formulation is to have a value and a range around (or above, or below) such a value.

In all these settings, it is often possible to express the instability by a bounding range of values. As another example, we may have a default cost of 10 for each piece of a material, with a range from 5 to 15 containing all possible foreseen alternatives costs.

Even though costs are unstable, we would still like to reason and perform inference on the given "default" optimization problem. This is possible in some cases, as we will see in this paper.

Given constraint optimization problems where some of the costs are tagged as unstable, we define several notions of desirable solutions, that take into account not only cost-optimality, but also a form of robustness (of the optimality status, or of the cost) with respect to the uncertainty present in the problem. For example, we could desire solutions that are cost-optimal and that remain optimal even if the unstable costs change. In other scenarios, it could instead be important to find solutions that are cost-optimal and whose cost does not increase if the unstable costs change.

Some of the considered notions will yield sets of solutions that can possibly be empty, while others (usually the least attractive) will always have at least a single element. For each of the notions of optimality, we provide an algorithm to find solutions according to that criterion, and we study their properties.

In addition to the notion of instability, we also accomodate the dichotomy of having some uncontrollable variables, whose value cannot be decided by the agent, but will be decided by Nature or by some other agent. This yields an orthogonal form of uncertainty often found in scheduling or temporal problems [12], where the occurrence of certain events can be decided only by others. For example, in scheduling the activities of a satellite taking pictures of Earth, we may have to schedule in advance the best times for taking some pictures of an area without knowing the local weather conditions (that heavily impacts on the quality of the pictures), which is decided by Nature.

To handle the uncontrollable variables, we adopt a variant of classical variable elimination, where we act pessimistically. This allows processing the uncontrollable variables first, and then working on the controllable part ensuring that the resulting inferences are safe with respect to the uncontrollable part.

Interestingly, no matter what notion of optimality we use, the complexity of reasoning with the unstable and uncontrollable problems considered in this paper do not increase the overall worst case complexity (for complete algorithms). In particular, if the default problem belongs to a tractable class, even its unstable and/or uncontrollable version is tractable.

Issues related to those considered in this paper have been studied also in open CSPs [6] and interactive CSPs [9]. However, in such frameworks the uncertainty is in the form of missing domain values, and not unstable costs. Also, in dynamic CSPs [3], variables, domains, and constraints may change over time. However, no preference range is given, and there are no uncontrollable variables. Preference ranges are considered in [14], however no default value is given in the various preference ranges. In [7] only uncontrollable variables are considered, but no imprecise ranges are given. In [8] some preferences are missing (thus there are no default value nor preference ranges) and the focus is on preference elicitation to obtain the so-called necessarily optimal solutions (called O-ROB in this paper). A similar setting can be found in [13] for hard CSPs.

2   Background: Soft Constraints for Optimization Problems

A soft constraint [1] is just a classical constraint [2] where each instantiation of its variables has an associated value from a (totally or partially ordered) set. This set has two operations, which makes it similar to a semiring, and is called a c-semiring.

More precisely, a c-semiring is a tuple $\langle A, +, \times, \mathbf{0}, \mathbf{1} \rangle$ containing a set of preferences A, a combination operation $\times$, that is useful to combine preferences, and an additive operator $+$ that induces a partial order $\leq$ over A. Such an ordering gives us a way to compare (some of the) tuples of values and constraints. In fact, when we have $a \leq_S b$, we will say that b is better than a. Thus, $\mathbf{0}$ is the worst value and $\mathbf{1}$ is the best one. The combination operator is intensive, that is, $\forall a, b \in A$, $a \times b \leq_S a, b$.

A c-semiring $\langle A, +, \times, \mathbf{0}, \mathbf{1} \rangle$ is said to be strictly monotonic iff the combination operator $\times$ is strictly monotonic, i.e., for every $a, b \in A$, if $a < b$ then, for every $c \in A$, $a \times c < b \times c$.

Given a set of variables V with finite domain D, and a c-semiring $\langle A, +, \times, 0, 1 \rangle$, a soft constraint is a pair $\langle def, con \rangle$ where $con \subseteq V$ is the scope of the constraint and $def : D^{|con|} \longrightarrow A$ is the preference function of the constraint associating to each tuple of assignments to the variables in con either a preference value ranging between $\mathbf{0}$ and $\mathbf{1}$. A soft constraint problem (SCSP) is a triple $\langle C, V, D \rangle$, where C is a set of soft constraints over the variables in V with domain D.

Given two constraints $c_1 = \langle def_1, con_1 \rangle$ and $c_2 = \langle def_2, con_2 \rangle$, their *combination* $c_1 \otimes c_2$ is the constraint $\langle def, con \rangle$ defined by $con = con_1 \cup con_2$ and $def(t) = def_1(t \downarrow_{con_1}^{con}) \times def_2(t \downarrow_{con_2}^{con})$[1]. In words, combining two constraints means building a new constraint which involves all the variables of the original ones and which associates to each tuple of domain values for its variables a specific semiring element. Such an element is obtained by multiplying the elements associated by the original constraints to the appropriate subtuples.

It may be useful to eliminate some variables from a constraint, using a notion of projection. Given a subset of variables $I \subseteq V$, and a soft constraint $c = \langle def, con \rangle$, the *projection* of c over I, written $c \Downarrow_I$, is a new soft constraint $\langle def', con' \rangle$, where $con' = con \cap I$ and $def(t') = \sum_{\{t \mid t \downarrow_{con'} = t'\}} def(t)$. In particular, the scope, con', of the projection constraint contains the variables that con and I have in common, and thus $con' \subseteq con$. Moreover, the preference associated to each assignment to the variables in con', denoted with t', is the highest ($\sum$ is the additive operator of the c-semiring) among the preferences associated by def to any completion of t', t, to an assignment to con.

Many known classes of satisfaction or optimization problems can be cast in this formalism. For example, a classical CSP is just an SCSP where the chosen c-semiring is: $S_{CSP} = \langle \{false, true\}, \vee, \wedge, false, true \rangle$. In fact, constraints can only be either satisfied (true) or violated (false), the logical and models the fact that we want all the constraints to be satisfied, and the logical or models the fact that we prefer true to false.

Fuzzy CSPs can be modeled in the SCSP framework by choosing the c-semiring: $S_{FCSP} = \langle [0, 1], max, min, 0, 1 \rangle$. This means that preferences are values between 0 and 1. The max operator shows that we prefer higher values to lower ones. The min operator says that, when we combine preferences of several constraints, we take the lowest value. Thus in fuzzy CSPs we want to maximize the minimum preference. This is a pessimistic approach to preference handling, that works well in application domains where one needs to be very cautious, such as medical or space applications.

For weighted CSPs, the semiring is $S_{WCSP} = \langle \Re^+, min, +, +\infty, 0 \rangle$. Here preferences are interpreted as costs from 0 to $+\infty$, which are combined with the sum and compared with min. Thus the optimization criterion is to minimize the sum of the costs.

Given an assignment s to all the variables of an SCSP $P = \langle C, V, D \rangle$, we denote by $pref(s, P)$ the preference of s in P, defined as $pref(s, P) = \prod_{\langle def, con \rangle \in C} def(s_{\downarrow con})$. In words, it is obtained by taking the combination of the preferences associated to the sub-tuples corresponding to the solution in the constraints. A complete assignment of values to all the variables is an *optimal solution* if its preference is the best one w.r.t. the ordering induced by the additive operator. Thus, if we are working with fuzzy CSPs, its preference value must be the highest one, and if we are working with weighted CSPs, its cost must be the lowest. Given an SCSP P, we denote with $Opt(P)$ the set of all the optimal solutions of P.

---

[1] By $t \downarrow_Y^X$ we mean the subtuple obtained by projecting the tuple t (defined over the set of variables X) over the set of variables $Y \subseteq X$.

Semiring-based soft constraints model optimization problems by using preferences in the constraints and combining them via the semiring combination operator. This induces an ordering over the solutions of the problem, which can be seen as an objective function. Thus soft constraints can model all objective functions that are decomposable over the topology of the problem.

Techniques used to find optimal solutions of constraint optimization problems can be divided into search-based schemes and inference-based schemes [2]. The most common search-based algorithm for constraint optimization is *Branch and Bound* (BB) [2,11]. On the other hand, a very general inference-based algorithm is *Bucket elimination* (BE) [5], which may be seen as an extension of adaptive consistency [2] to optimization problems. Given a linear order over the variables, in the bucket processing phase, each variable is considered, in one direction of the order, and it is removed by projecting the combination of the constraints involving it over all the other variables in such constraints. After all the variables (but one) have been eliminated, in the forward phase, the variables can be assigned, following the other direction of the order, and an optimal solution can be found in polynomial time.

Both techniques have an exponential worst time case complexity. BE, in contrast with BB, also needs possibly exponential space, but it can exploit the graph structure of the problem. For some structures, such as problems with tree-shaped constraint graphs, optimal solutions can be found in polynomial time [2]. Recent extension of Branch and Bound strategies that explore the AND/OR search space of a graphical model were shown to allow similar complexity bounds to inference-based schemes [4,10].

3   Unstable Optimization Problems

We define unstable SCSPs as SCSPs where there may be some unstable preferences. Such preferences are specified by a default value d plus an an interval $[l, u]$, that contains all possible values that can replace the default value d.

Definition 1 (unstable soft constraint). *Given a set of variables V with finite domain D, and a c-semiring $S = \langle A, +, \times, 0, 1 \rangle$, an unstable soft constraint is a pair $\langle f, con \rangle$ where $con \subseteq V$ is the scope of the constraint and the preference function of the constraint is $f : D^{|con|} \longrightarrow A \times I$, s.t. $t \mapsto (d, [l, u])$, where I is a set of all the intervals of values in A and $l \leq_S d \leq_S u$. All tuples mapped into $(d, [l, u])$ where $l <_S u$ (resp., $l = u$) are called unstable (resp., stable) tuples and their preference d is called an unstable (resp., stable) preference.*

In what follows, when there is a stable preference, instead of writing $(d, [d, d])$ we will simply write d. Also, when it is clear from the context, we will omit the semiring name and we will write $\leq$ instead of $\leq_S$. Notice that $\leq_S$ is not always the usual $\leq$ over naturals or reals. In fact, if we are dealing with costs, where higher means worst in the semiring, we have that $c_1 \leq_S c_2$ when $c_2 \leq c_1$.

As an example of an unstable constraint using the fuzzy c-semiring $\langle [0, 1], max, min, 0, 1 \rangle$, consider $V = \{X, Y\}$, $D = \{a, b\}$, $con = V$, and preference function

$f(X = a, Y = a) = 0.1$, $f(X = a, Y = b) = (0.5, [0.3, 0.6])$, $f(X = b, Y = a) = 0.6$, $f(X = b, Y = b) = (0.7, [0.5, 1])$.

Instead, as an example of an unstable constraint using the weighted c-semiring $\langle \Re^+, min, +, +\infty, 0 \rangle$, we can consider a constraint with $con = V$ and cost function $g(X = a, Y = a) = 100$, $g(X = a, Y = b) = (50, [60, 20])$, $g(X = b, Y = a) = 80$, $g(X = b, Y = b) = (30, [50, 10])$. Notice that according to the ordering induced by the weighted c-semiring, we have, for example, $50 < 10$, since 10 is better than 50 in the weighted semiring.

Definition 2 (unstable SCSP). *An unstable SCSP (USCSP) is a tuple $\langle S, V, D, C \rangle$, where V is a set of variables with domain D and C is a set of unstable soft constraints over the variables in V over the c-semiring S.*

An USCSP where all the preferences are stable corresponds to an SCSP.

Definition 3 (solution). *A solution of an USCSP is an assignment to all its variables.*

We now introduce our running example. Consider the problem related to building a piece of furniture with some iron. Assume that for iron we may have high, medium, or bad quality, with costs 50, 30, and 20. We also assume that the processing time for the piece of furniture is 2 or 3 days, and that the processing cost depends on the quality of the iron and on how many work days are needed. This problem can be modelled by an USCSP over the weighted c-semiring with:

- two variables Q_i and T representing the quality of the iron and the processing time, with domains $D(Q_i) = \{b, m, h\}$ and $D(T) = \{2, 3\}$;
- an unstable soft constraint on Q_i with cost function f_i defined by $f_i(b) = 20$, $f_i(m) = 30$, and $f_i(h) = 50$;
- an unstable soft constraint on Q_i and T, with cost function f defined by $f(h, 2) = 10$, $f(h, 3) = 20$, $f(m, 2) = (30, [60, 5])$, $f(m, 3) = (35, [100, 20])$, $f(b, 2) = 80$, $f(b, 3) = 100$. Thus, for example, if the iron is of bad quality, the processing cost is 80 if the work is done in 2 days, and 100 is it is done in 3 days. Also, if the quality is medium, and the work is done in 2 days, we expect the processing cost to be 30. However, this value may change in the range $[60, 5]$. Similarly, if the work is done in 3 days, we expect the processing cost to be 35, but it can change in the range $[100, 20]$. A solution is, for example, $(Q_i = h, T = 3)$: high quality iron is used and three days of work are needed.

Clearly, not all solutions are equally desirable. In order to discriminate among them, we will define some optimality notions for USCSPs, as well as algorithms to handle them. To do this, we start by giving some basic notions which will be useful in what follows.

Given an USCSP P, a *scenario* of P is an SCSP obtained from P by replacing every unstable preference with a value in its range. $SC(P)$ denotes the set of all possible scenarios of P.

In terms of defining the notions of optimality, a special role will be played by the *default scenario*, denoted by P_d, where only default values are considered. Such a scenario represents the problem given by the user when instability is ignored. Moreover, it will be useful to consider the *worst scenario*, denoted by P_l, where only the worst elements in the ranges are considered, and the *best scenario*, denoted by P_u, where only the best elements in the ranges are considered. We will also denote the preference value of the optimal solutions of P_d, P_l, and P_u, by, respectively, $pref_d$, $pref_l$, and $pref_u$.

In the running example, we have that $pref_d = 60$, obtained by solutions $(Q_i = m, T = 2)$ and $(Q_i = h, T = 2)$. Also, $pref_u = 35$, obtained by solution $(Q_i = m, T = 2)$. Finally, $pref_l = 60$, obtained by $(Q_i = h, T = 2)$.

4   Optimal and Optimality-Robust Solutions (O-ROB)

The first kind of solutions that we consider are optimal solutions that are robust w.r.t. optimality. This means that their status of being optimal does not change, regardless of any variation of the unstable preference values within their ranges.

Such a notion of optimality is useful when it is necessary to adopt a safe attitude: we want our decision to be optimal no matter what happens to the unstable parts of the problem.

Definition 4 (O-ROB). *Given an USCSP P, a solution is in O-ROB(P)) iff*

– *it is optimal in P_d, and*
– *it is optimal in all other $P' \in SC(P)$.*

Note that these solutions were called necessarily optimal in [8], that considers problems where every interval is the largest one, and there were no default values. While considering any range adds expressiveness, the presence of default values is not important for this notion of optimality. In fact, the above definition could easily be replaced by an equivalent one (more compact but less easy to relate to the problem definition) where we only require s to be optimal in all $P' \in SC(P)$. In fact, P_d is just one of the problems in $SC(P)$.

Proposition 1. *Given an USCSP P, the set O-ROB(P) may be empty.*

In fact, in the running example, the optimal solutions of P_d are $(Q_i = m, T = 2)$, and $(Q_i = h, T = 2)$. However, $(Q_i = m, T = 2)$ is not optimal in P_l, and $(Q_i = h, T = 2)$ is not optimal in P_u. Thus O-ROB(P) is empty.

Algorithm 1 shows a procedure to find solutions in O-ROB(P).

Find-OROB takes in input an USCSP P and returns either a solution in O-ROB(P), or nil. To do this, it computes an optimal solution s_l of P_l and an optimal solution s_u of P_u, with preferences $pref_l$ and $pref_u$. This can be done via any of the solving techniques for SCSPs (denoted with *Solve* in the pseudocode). Then, if $pref_l = pref_u$, Find-OROB returns the optimal solution of P_l, otherwise it returns *nil*. It is possible to show that Algorithm Find-OROB is sound, but not complete. Thus, if it returns a solution, it is in O-ROB(P). If instead it returns nil, this does not necessarily mean that O-ROB(P) is empty.

Algorithm 1. Find-OROB

Input: an USCSP P; **Output**: a solution or nil
$(s_l, pref_l) \leftarrow Solve(P_l); (s_u, pref_u) \leftarrow Solve(P_u)$
if $pref_l = pref_u$ **then**
$\quad \llcorner$ **return** s_l
else
$\quad \llcorner$ **return** nil

Theorem 1. *Given an USCSP P, if Find-OROB(P) = s_l, then $s_l \in$ O-ROB(P).*

Proof. Assume Find-OROB$(P) = s_l$. Due to the monotonicity of the combination operator off the semiring, the preference of s_l in any scenario can only be higher than, or equal to, its value $pref_l$ in P_l. Since $pref_l = pref_u$, this means that whatever preference values are assigned to unstable tuples within their ranges, the preference of s_l is always $pref_l$ and the preference of any other solution is never greater than $pref_l$. Thus s_l is optimal in every scenario, and therefore it is in O-ROB(P). Q.E.D.

If Find-OROB$(P) = nil$, consider a very simple USCSP P with one variable X with domain $\{a, b, c\}$ and with a unary unstable constraint over X with cost function $f(a) = (10, [20, 5]), f(b) = f(c) = 30$. It is easy to see that $pref_u = 5$, $pref_l = 20$, but $(X = a)$ is in O-ROB(P). Thus Find-OROB$(P) = nil$ but O-ROB(P) is not empty.

Notice that, if $pref_l = pref_u$, not every solution of P_u is in O-ROB(P), since there might be ways to set the unstable preferences that make that solution not optimal in some scenarios. This is why we can only take the solutions in P_l.

Algorithm Find-OROB is therefore sound and not complete. However, finding a solution in O-ROB(P) with algorithm Find-OROB requires just solving two optimization problems. In [8] it is shown that this approach is both sound and complete when we restrict the preference ranges to be all equal to the $[\mathbf{0}, \mathbf{1}]$ interval, where $\mathbf{0}$ and $\mathbf{1}$ are the worst and the best preference values, and $pref_l > \mathbf{0}$.

5 Optimal and Preference-Robust Solutions (P-ROB)

Another kind of optimal solutions that we consider are those that are robust w.r.t. their preferences. That is, solutions that are optimal in the default scenario, and that do not require additional cost if the scenario changes. However, they could loose their optimality status if the scenario changes.

This notion is useful when we act under severe cost restrictions: we would like our decisions to be optimal at least in the default scenario, and be sure that no additional cost is needed if the unstable costs turn out to be different from the default ones.

Definition 5 (P-ROB). *Given an USCSP P, a solution s is in P-ROB(P) iff*

- *it is optimal in P_d and*
- *$\forall P' \in SC(P), pref(s, P') \geq pref(s, P_d) = pref_d$.*

In words, a solution is in P-ROB(P) iff it is optimal in P_d and its preference ($pref_d$) may only improve if the scenario changes. Such solutions are interesting whenever the optimal cost of the default problem is attractive, and we want to make sure that in all other scenarios no additional cost will be required.

In the running example, among the optimal solutions of P_d, only ($Q_i = h, T = 2$) is in P-ROB(P).

Proposition 2. *Given an USCSP P, P-ROB(P) may be empty.*

In fact, if we consider the USCSP R obtained from the USCSP defined in the running example by changing the stable preference of ($Q_i = h, T = 2$) from 10 to 20, the only optimal solution of R_d is ($Q_i = m, T = 2$), but its preference worsens in R_l. Thus P-ROB(R) is empty.

Algorithm 2 shows a method which, given in input a USCSP P, returns a solution in P-ROB(P) if there is any.

Algorithm 2. Find-PROB

Input: an USCSP P; **Output**: a solution, or nil
$(s_l, pref_l) \leftarrow Solve(P_l)$; $(s_d, pref_d) \leftarrow Solve(P_d)$
if $pref_d = pref_l$ **then**
   ⌊ **return** s_l
else
   ⌊ **return** *nil*

We will now prove that Algorithm Find-PROB is sound and complete.

Theorem 2. *Given an USCSP P, if Find-PROB(P)=s then $s \in$ P-ROB(P) and if Find-PROB(P)=nil then P-ROB(P)=$\emptyset$.*

Proof. Assume Find-PROB(P) returns a solution s. This happens if and only if s is optimal in P_l. Since $pref_d = pref_l$, and due to monotonicity of the multiplicative operator of the c-semiring, s is optimal also in P_d. Again due to monotonicity, $\forall P' \in SC(P)$, $pref(s, P_d) = pref(s, P_l) \leq pref(s, P')$. Thus, by definition, $s \in$ P-ROB(P).

If Find-PROB(P) returns *nil* then $pref_d > pref_l$, that is, for any optimal solution s of P_d, $pref(s, P_d) > pref(s, P_l)$. This means that P-ROB($P$)=$\emptyset$. Q.E.D.

To find a solution in P-ROB(P) with algorithm Find-PROB, it is enough to solve two optimization problems. This was true also for Find-OROB, but Find-PROB is both sound and complete.

6 Optimality-Robust and Preference-Robust Solutions (OP-ROB)

A solution is robust w.r.t. to both optimality and preferences if it is optimal in the default scenario, and both its optimality status and its cost do not worsen if the scenario changes. This is the strongest and most desirable notion.

This notion of optimality is useful when we have both cost restrictions and stringent user requirements: the user wants a solutions which is optimal no matter what, and the company wants to make sure that there is no additional costs if a scenario different from the default one occurs.

Definition 6 (OP-ROB). *Given a USCSP P, a solution $s \in$ OP-ROB(P) iff*

- *it is optimal in P_d,*
- *it is optimal in all other $P' \in SC(P)$, and*
- $\forall P' \in SC(P), pref(P', s) \geq pref(P_d, s)$.

It is easy to see that a solution is in OP-ROB(P) iff it is in O-ROB(P) $\cap$ P-ROB(P). In the running example, since we have shown in Section 4 that OROB(P) is empty, also OP-ROB(P) is empty.

Proposition 3. *Given an USCSP P, the set OP-ROB(P) may be empty.*

This follows immediately from the fact that O-ROB(P) $\cap$ P-ROB(P) = OP-ROB(P) and that both O-ROB(P) and P-ROB(P) may be empty.

To find such solutions, we combine the two procedures Find-OROB and Find-PROB as shown in Algorithm 3.

Algorithm 3. Find-OPROB

Input: an USCSP P; **Output**: a solution s, or nil
$(s_d, pref_d) \leftarrow Solve(P_d)$; $(s_l, pref_l) \leftarrow Solve(P_l)$; $(s_u, pref_u) \leftarrow Solve(P_u)$
if $pref_d = pref_l = pref_u$ **then**
  $\llcorner$ **return** s_l
else
  $\llcorner$ **return** nil

Algorithm Find-OPROB, given in input an USCSP P checks if $pref_d = pref_u = pref_l$. If this is so, it returns an optimal solution of P_l, otherwise it returns nil. This method is sound but not complete, as shown in the following theorem.

Theorem 3. *Given an USCSP P, if Find-OPROB(P) $= s$, then $s \in$ OP-ROB(P). If Find-OROB(P)=nil, then OP-ROB(P) might be not empty.*

Proof. If Find-OPROB(P)=s, $pref_d = pref_u = pref_l$. By Theorem 1, $s \in$ O-ROB(P). Also, by Theorem 2, $s \in$ P-ROB(P). Thus $s \in$ O-ROB(P) $\cap$ P-ROB(P) = OP-ROB(P).

In order to show that, when Find-OROB(P)=nil, OP-ROB(P) might be not empty, let us consider an USCSP P with one variable X with domain $\{a, b, c\}$ and with a unary unstable constraint over X with cost function $f(a) = (20, [20, 5])$, $f(b) = f(c) = 30$. It is easy to see that $pref_u = 5$, $pref_l = pref_d = 20$, but $(X = a)$ is in OP-ROB(P). Thus Find-OROB(P)=nil but OP-ROB(P) is not empty. Q.E.D.

Finding a solution in OP-ROB(P) using algorithm Find-OPROB amounts to solving three SCSPs.

7   The Best Preference-Robust Solutions (Best-ROB)

Solutions in P-ROB(P) are optimal in the default scenario and their cost never increases if the scenario changes. If the main focus is avoiding additional costs when the scenarios changes, rather than the optimality in the default scenario, we can relax the first requirement. The set of solutions of this kind will be denoted by Best-ROB(P). A solution s is in Best-ROB(P) if its cost in P_d can only decrease by changing the scenario. Also, among the solutions with such a property, it is the one with lowest cost in P_d.

Thus, a solution in Best-ROB(P) could be non-optimal in the default scenario. However, there is no better solution in Best-ROB(P) whose cost does not increase in some other scenarios.

Solutions of this kind are useful, for example, when budget limitations guide the operations of a company more than solutions quality. In fact, such solutions assure that no additional cost is needed, although they may sacrifice solution optimality to achieve this.

Definition 7 (Best-ROB). *Given an USCSP P, a solution $s \in$ Best-ROB(P) iff*

- *$s \in F = \{s|\ pref(s, P_d) > \mathbf{0}$ and $\forall P' \in SC(P),\ pref(s, P') \geq pref(s, P_d)\}$ and*
- *$\forall s' \in F,\ pref(s, P_d) \geq pref(s', P_d)$.*

Notice that, in general, if P-ROB(P) $\neq \emptyset$, then Best-ROB(P) = P-ROB(P). This is the case of our running example, where both sets contain only solution ($Q_i = h, T = 2$).

Proposition 4. *Given an USCSP P, the set Best-ROB(P) may be empty.*

To see this, let us consider the USCSP P with one variable X with domain $\{a, b, c\}$ and with a unary unstable constraint over X with cost function $f(a) = f(b) = f(c) = (10, [20, 5])$. In such a case, all solutions have a cost in P_l which is strictly higher than that in P_d. Thus Best-ROB(P)=$\emptyset$.

Algorithm 4 shows the procedure for the strictly monotonic case which uses an SCSP denoted with P_{fix}. P_{fix} is obtained from the USCSP P in input by just fixing the unstable preferences as follows: for each unstable preference $(d, [l, u])$ in P, we put in P_{fix}, $\mathbf{0}$ if $l < d$, and d otherwise. The intuition behind the construction of P_{fix} is to forbid those tuples associated to preferences that may worsen w.r.t. their default values when the scenario changes.

Find-BestROBm checks if SCSP P_{fix} has a solution with preference strictly better than $\mathbf{0}$. If so, it returns an optimal solution of P_{fix}, otherwise it returns *nil*. This algorithm to find solutions in Best-ROB(P) is both sound and complete.

Theorem 4. *Given an USCSP P over a strictly monotonic c-semiring, if Find-BestROBm(P) = s, s $\in$ Best-ROB(P). Find-BestROBm(P)=nil iff Best-ROB(P)=$\emptyset$.*

Algorithm 4. Find-BestROBm

Input: an USCSP P with a strictly monotonic c-semiring; **Output**: a solution
s, or nil
$(s, p) \leftarrow Solve(P_{fix})$
if $p > \mathbf{0}$ **then**
└ **return** s
else
└ **return** nil

Proof. Find-BestROBm$(P) = s$ iff $p > \mathbf{0}$. We need to show that $Opt(P_{fix}) \subseteq$
Best-ROB(P). By construction of P_{fix}, and due to the strict monotonicity of the
combination operator, we have that, $\forall s' \in Opt(P_{fix})$, if $pref(s', P_{fix}) > \mathbf{0}$, then
$pref(s', P_{fix}) = pref(s', P_d)$ and, $\forall P' \in SC(P)$, $pref(s', P') \geq pref(s', P_d)$.
Thus, since $s \in Opt(P_{fix})$, it satisfies this property and there is no solution with
a higher preference in P_d satisfying it. This means that $s \in$Best-ROB(P).

Find-BestROBm$(P) = nil$ iff $p = \mathbf{0}$, that is, iff all solutions of P_{fix} have preference $\mathbf{0}$. Thus, for all s such that $pref(s, P_d) > \mathbf{0}$, $s \notin$ Best-ROB(P). The other
solutions, that have preference $\mathbf{0}$ in P_d, are not in Best-ROB(P) by definition.
Thus Best-ROB(P) $= \emptyset$. Q.E.D.

This theorem shows that algorithm Find-BestROBm is both sound and complete. Finding solutions in Best-ROB(P) using this algorithm (that is, when
the combination operator is strictly monotonic) amounts at solving one SCSP.
However, this algorithm works only when the combination operator is strictly
monotonic. It is possible to define a sound but not complete approach that works
for any c-semiring.

Consider applying BE to scenarios P_d and P_l using the same variable ordering.
At the end of the bucket processing phase [2] in both scenarios, we obtain two
new SCSPs, say P'_d and P'_l, where there are additional constraints and possibly
lower preferences in the old constraints. The first variable in the linear order,
say x, has each value, say a, in its domain associated with the highest preference
(or lowest cost) of a solution of the corresponding scenario where $x = a$.

We then check if there are values for x that have the same preference ·in
both P'_l and P'_d. If this is not the case, we are not able to say anything about
set Best-ROB(P). Otherwise, we pick among such values one, say a, with the
highest preference, say p. Using the forward step of BE applied to $x = a$ in P'_l,
assignment $x = a$ can be extended to a solution of P'_l (and thus of P_l) with
preference p. Due to the monotonicity of the combination operator and to the
fact that $x = a$ has the same preference in P'_l and P'_d, such a solution has the
same preference p also in P'_d (and thus in P_d). Moreover, there is no solution with
this property and a higher preference in P_d. This means that such a solution is
in Best-ROB(P).

This algorithm can always be used, but it is possibly not complete. It requires
to solve two SCSPs with BE to find a solution in Best-ROB(P).

8   The Most Preference-Robust Optimal Solutions (ROB-OPT)

Another way to be more tolerant with the conditions of P-ROB(P) is to relax the second requirement, that is, to maintain optimality in the default scenario but to allow for a decrease in the preference if the scenario changes. However, such a decrease should be the smallest possible in the worst scenario. This set of solution is called ROB-OPT(P).

Definition 8 (ROB-OPT). *Given an USCSP P, a solution $s \in ROB\text{-}OPT(P)$ iff*

- *it is optimal in P_d, and,*
- *for every other optimal solution of P_d, say s', $pref(s, P_l) \geq pref(s', P_l)$.*

In words, a solution is in ROB-OPT(P) if it is optimal in the default scenario and, among the solutions that are optimal in such a scenario, its preference value decreases the least in the worst scenario.

Contrarily to all the previous notions of optimality, this set always contains at least a solution. In general, if P-ROB(P) $\neq \emptyset$, P-ROB(P) = ROB-OPT(P). This is the case of our running example, where ROB-OPT(P) = $\{(Q_i = h, T = 2)\}$.

Proposition 5. *Given an USCSP P, ROB-OPT(P) is never empty.*

It follows immediately from the fact that the set of optimal solutions of P_d is never empty.

To find solutions in ROB-OPT(P), we propose a procedure based on defining a new SCSP. Given an USCSP P, defined over the c-semiring $S = \langle A, +, \times, 0, 1 \rangle$, we consider the SCSP P_{dl} with the same variables and constraint topology as P, and defined over the c-semiring $S' = \langle A \times A, lex(+, +), (\times, \times), (\mathbf{0}, \mathbf{0}), (\mathbf{1}, \mathbf{1}) \rangle$. In such a c-semiring, preferences are pairs which are combined by applying the combination operator to the corresponding components and which are ordered lexicographically with the first component being the most important one. In P_{dl}, each tuple associated with preference $(d, [l, u])$ in P is instead associated with preference (d, l). The intuition behind the definition of P_{dl} is that, by solving it, we find solutions which in the first place maximize the combination of the default preferences, and secondly maximize the combination of the lower bounds of the ranges of the unstable preferences.

Theorem 5. *Given an USCSP P and a solution $s = Find\text{-}ROBOPT(P)$, $s \in ROB\text{-}OPT(P)$.*

Algorithm 5. Find-ROBOPT

Input: an USCSP P; **Output:** a solution s
$(s, p) \leftarrow Solve(P_{dl})$
return s

Proof. Since Find-ROBOPT(P) always returns an optimal solution of P_{dl}, it is sufficient to prove that $Opt(P_{dl}) = $ ROB-OPT(P). We first show that $Opt(P_{dl}) \subseteq$ ROB-OPT(P). Notice that, by construction of P_{dl}, if a solution s has preference (d, l) in P_{dl}, then in P_d it has preference d and l in P_l. Given a solution $s \in Opt(P_{dl})$ with preference (d, l), for every other solution s' of P_{dl} with preference (d', l'), it must be that $d \geq d'$. Thus, $s \in Opt(P_d)$. Moreover, for every other solution s'' with preference (d, l''), it must be that $l \geq l''$. Thus $s \in$ ROB-OPT(P).

Next we show that ROB-OPT(P) $\subseteq Opt(P_{dl})$. If $s \in$ ROB-OPT(P), it is optimal in P_d and thus, in P_{dl}, it is among the solutions with a maximal first component. Moreover, among the optimal solutions of P_d, s is one of the solutions with the highest preference in P_l. This means that among those with the maximal first component in P_{dl}, it has the highest second component. This corresponds to being undominated w.r.t. the ordering induced by $lex(+, +)$. Q.E.D.

9 USCSPS with Uncontrollable Variables

In unstable SCSPs all the variables are decided by the deciding agent. The only form of uncertainty is the presence of the preference ranges around the default values. However, in many real-life settings, there are also variables which are *uncontrollable*, that is, their value cannot be chosen by the deciding agent. Such a value will be decided by some other agent or by Nature. Typical examples are times of events related to weather changes. For example, we don't know when the clouds will disappear.

We will now consider the presence of some of this kind of variables in an unstable SCSP. In this generalized setting, the variables V of the problem will be partitioned in two sets, namely V_c and V_u, containing, respectively, the controllable and the uncontrollable variables. In this paper we assume to have no information on the uncontrollable variables besides their domain values. For example, we don't have a probability, and not even a possibility, distribution over such a domain.

There are many ways to reason with uncontrollable variables, which depend on the attitude to risk of the agent. Examples are the notions of strong, weak, and dynamic controllability in temporal reasoning [12]. Here we adopt a pessimistic approach (which follows the same principle as for strong controllability), where we want to make sure that the cost of the decision we take over the controllable part of the problem is guaranteed not to increase when the uncontrollable variables are instantiated.

To achieve this, we first *eliminate* the uncontrollable variables one at a time in a linear order. For each variable v in V_u, we consider all constraints $c_1, \ldots, c_n$ connecting v to other variables, say $v_1, \ldots, v_m$, and we build a new constraint c connecting $v_1, \ldots, v_m$, whose preference function f is defined as follows: $f(d_1, \ldots, d_m) = \Pi_{d \in D(v)} \Pi_{i=1}^{n} l_i(t_i, d)$, where d_i is the subtuple of $(d_1, \ldots, d_m)$ involving only values for the variables in $con(c_i)$, and $l_i(t_i, d)$ is the lower element of the range associated to tuple (t_i, d) by the preference function of constraint

c_i. In words, we associate to tuple $(d_1, \ldots, d_m)$ the greatest lower bound of the preferences that can be obtained by extending this tuple to any value in the domain of v.

Notice that this procedure can be seen as a variant of BE where we take the worst case rather than the best one (as is done in the projection step).

Uncontrollable variables are eliminated one at a time, until only controllable variables are left. Given an USCSP P with uncontrollable variables, we denote by $cont(P)$ the resulting USCSP obtained by applying this procedure to P.

Theorem 6. *Consider an USCSP P with controllable variables V_c and uncontrollable variables V_u. For any assignment s to the variables in V_c, and any assignment s' to the variables in V_u, $pref(s, cont(P)) \leq_S pref((s, s'), P)$, where S is the c-semiring over which P is defined.*

Proof. It follows by monotonicity and intensivity of the $\times$ operator of the semiring. Q.E.D.

We can therefore reason on an unstable SCSP P with uncontrollable variables by first eliminating all uncontrollable variables, thus obtaining $cont(P)$, and then by reasoning on $cont(P)$ according to any one of the optimality/robustness criteria defined in the previous sections. No matter what solution we end up with, we are sure that no additional cost will be needed when the values of the uncontrollable variables will be known. For example, if we choose to use O-ROB, we find a solution of the controllable part which is optimal for the default scenario and remains optimal even if the unstable preferences change, and whose preference level cannot decrease because of how Nature decides to instantiate the uncontrollable variables.

10   Final Considerations and Future Work

For most of the notions of optimality considered in this paper, for which we give sound and complete algorithms, finding an optimal solution for an unstable SCSP P requires solving at most three SCSPs. This means that, to handle the kind of uncertainty modelled by preference ranges and default values, we don't change the complexity class. In particular, for example, if the default problem belongs to a tractable SCSP class, this is also true for any unstable SCSP with the same topology.

We did not consider probability distributions over the ranges of the possible values for the unstable costs. We believe there are several application domains where probabilistic reasoning is not suitable, and one would rather prefer to reason with exact, although unstable, information. However, we also envision domains where it makes sense to consider the expected utility of a solution or a scenario, and to take decisions based on such concepts. We plan to study this adaptation of our work.

We also plan to implement the algorithms to obtain the several notions of optimal solutions defined in this paper, and also to test experimentally how many times those notions that may return an empty set actually do this. While these notions seems to be less appealing because there may be none of them, it may be that in certain application domains, or in classes of problems with a certain structure, there are always some of them.

References

1. Bistarelli, S., Montanari, U., Rossi, F.: Semiring-based constraint solving and optimization. Journal of the ACM 44(2), 201–236 (1997)
2. Dechter, R.: Constraint processing. Morgan Kaufmann, San Francisco (2003)
3. Dechter, R., Dechter, A.: Belief maintenance in dynamic constraint networks. In: AAAI, pp. 37–42 (1988)
4. Dechter, R., Mateescu, R.: And/or search spaces for graphical models. AI Journal 171(2-3), 73–106 (2007)
5. Dechter, R.: Bucket elimination: A unifying framework for reasoning. AI Journal 113(1-2), 41–85 (1999)
6. Faltings, B., Macho-Gonzalez, S.: Open constraint programming. AI Journal 161(1-2), 181–208 (2005)
7. Fargier, H., Lang, J., Schiex, T.: Mixed constraint satisfaction: a framework for decision problems under incomplete knowledge. In: Proceedings of the 13th National Conference on Artificial Intelligence (AAAI 1996), vol. 1, pp. 175–180. AAAI Press, Menlo Park (1996)
8. Gelain, M., Pini, M.S., Rossi, F., Venable, K.B.: Dealing with incomplete preferences in soft constraint problems. In: Bessière, C. (ed.) CP 2007. LNCS, vol. 4741, pp. 286–300. Springer, Heidelberg (2007)
9. Lamma, E., Mello, P., Milano, M., Cucchiara, R., Gavanelli, M., Piccardi, M.: Constraint propagation and value acquisition: Why we should do it interactively. In: IJCAI, pp. 468–477 (1999)
10. Mateescu, R., Dechter, R.: A comparison of time-space schemes for graphical models. In: Proc. IJCAI 2007, pp. 2346–2352. Morgan Kaufmann, San Francisco (2007)
11. Rossi, F., Van Beek, P., Walsh, T. (eds.): Handbook of Constraint Programming. Elsevier, Amsterdam (2006)
12. Vidal, T., Fargier, H.: Handling contigency in temporal constraint networks. JETAI 11(1), 23–45 (1999)
13. Wilson, N., Grimes, D., Freuder, E.C.: A cost-based model and algorithms for interleaving solving and elicitation of csps. In: Bessière, C. (ed.) CP 2007. LNCS, vol. 4741, pp. 666–680. Springer, Heidelberg (2007)
14. Yorke-Smith, N., Gervet, C.: Certainty closure: A framework for reliable constraint reasoning with uncertainty. In: Rossi, F. (ed.) CP 2003. LNCS, vol. 2833, pp. 769–783. Springer, Heidelberg (2003)

IDB-ADOPT:
A Depth-First Search DCOP Algorithm[*]

William Yeoh[1], Ariel Felner[2], and Sven Koenig[1]

[1] Computer Science
University of Southern California
Los Angeles, CA 90089-0781, USA
{wyeoh,skoenig}@usc.edu
[2] Information Systems Engineering
Ben-Gurion University of the Negev
Beer-Sheva, 85104, Israel
felner@bgu.ac.il

Abstract. Many agent coordination problems can be modeled as distributed constraint optimization (DCOP) problems. ADOPT is an asynchronous and distributed search algorithm that is able to solve DCOP problems optimally. In this paper, we introduce Iterative Decreasing Bound ADOPT (IDB-ADOPT), a modification of ADOPT that changes the search strategy of ADOPT from performing one best-first search to performing a series of depth-first searches. Each depth-first search is provided with a bound, initially a large integer, and returns the first solution whose cost is smaller than or equal to the bound. The bound is then reduced to the cost of this solution minus one and the process repeats. If there is no solution whose cost is smaller than or equal to the bound, it returns a cost-minimal solution. Thus, IDB-ADOPT is an anytime algorithm that solves DCOP problems with integer costs optimally. Our experimental results for graph coloring problems show that IDB-ADOPT runs faster (that is, needs fewer cycles) than ADOPT on large DCOP problems, with savings of up to one order of magnitude.

Keywords: ADOPT, DCOP, Distributed Constraint Optimization, Distributed Search Algorithms.

1 Introduction

Many agent coordination problems can be modeled as distributed constraint optimization (DCOP) problems, including the scheduling of meetings [8], the coordination of unmanned aerial vehicles [14], and the allocation of targets to

[*] This research was done while Ariel Felner spent his sabbatical at the University of Southern California, visiting Sven Koenig. This research has been partly supported by an NSF award to Sven Koenig under contract IIS-0350584. The views and conclusions contained in this document are those of the authors and should not be interpreted as representing the official policies, either expressed or implied of the sponsoring organizations, agencies, companies or the U.S. government.

A. Oddi, F. Fages, and F. Rossi (Eds.): CSCLP 2008, LNAI 5655, pp. 132–146, 2009.
© Springer-Verlag Berlin Heidelberg 2009

sensors in sensor networks [7,10,13]. Unfortunately, solving DCOP problems optimally is NP-hard. A variety of search algorithms have therefore been developed to solve DCOP problems as fast as possible to scale up to real-world domains [9,10,12,3]. ADOPT (Asynchronous Distributed Constraint Optimization) [10] is one of the pioneering DCOP algorithms and currently probably the most extended one [11,4,2]. It is an asynchronous and distributed best-first search algorithm that only needs a bounded amount of memory at each vertex and is able to solve DCOP problems optimally. Researchers have recently scaled up ADOPT by one order of magnitude by providing it with informed heuristics that focus its search [1]. However, its runtime is still large for realistic DCOP problems and it therefore needs to get scaled up further. In particular, in the original ADOPT, each vertex can switch back and forth between different values and then has to redo many searches since the results from the previous searches have already been purged from memory due to its memory limitations. In this paper, we address this problem of ADOPT by introducing Iterative Decreasing Bound ADOPT (IDB-ADOPT), a modification of ADOPT that changes the search strategy of ADOPT from performing one best-first search to performing a series of depth-first searches, where vertices do not switch back and forth between different values. IDB-ADOPT is motivated by insights from heuristic search that depth-first searches can outperform best-first searches in combinatorial domains with search trees whose depths are bounded [15], and DCOP problems are such domains. Each depth-first search of IDB-ADOPT is provided with a bound, initially a large integer, and returns the first solution whose cost is smaller than or equal to the bound. The bound is then reduced to the cost of this solution minus one. If there is no solution whose cost is smaller than or equal to the bound, it returns a cost-minimal solution. Thus, IDB-ADOPT is an anytime algorithm that solves DCOP problems with integer costs optimally. Our experimental results for graph coloring problems show that IDB-ADOPT runs faster than ADOPT on large DCOP problems, with savings of up to one order of magnitude.

2 DCOP Problems

Distributed constraint optimization (DCOP) problems model agent coordination problems as constraint optimization problems on *constraint graphs*. Each vertex

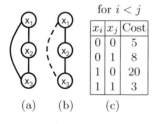

(a) (b) (c)

Fig. 1. Example DCOP problem

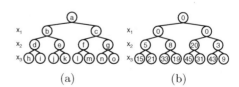

(a) (b)

Fig. 2. Search Tree

of a constraint graph represents an agent (sometimes referred to as variable) and can *take on* a value from a given set (its domain). Edges denote constraints. The cost of a constraint depends on the values of the vertex endpoints of the corresponding edge, given by a table. We assume in this paper that all costs are non-negative integers. An assignment of values to all vertices is a (complete) *solution*. The *cost* of the solution is the sum of the costs of the constraints. One wants to find a cost-minimal solution. As an example, Figure 1 (left) shows a simple DCOP problem with three vertices, x_1, x_2 and x_3, that each can take on the values zero or one. There are three constraints, whose costs are specified by the tables. For example, there is one constraint between vertices x_1 and x_2. If both vertices take on value zero, then the cost of the constraint is five. The cost of the solution $x_1 := 1$, $x_2 := 1$ and $x_3 := 1$ is nine and cost-minimal.

3   ADOPT

We now give a slightly simplified description of ADOPT that makes the search principle behind it easy to understand and is sufficient for our purposes. The reader is referred to the original paper [10] for a full step-by-step description of DCOP problems, ADOPT and the message passing mechanism used by ADOPT. ADOPT basically operates as follows: In a preprocessing step, ADOPT transforms the constraint graph into a *constraint tree* with the property that constraints exist only between a vertex and its ancestors and/or descendants. To simplify our description further, we assume that every vertex has at most one child in the constraint tree. In other words, the constraint tree is a chain, which is the case for our example DCOP problem. Figure 1 (center) shows one possible constraint tree for our example DCOP problem. ADOPT then performs a search as will be described next.

3.1   Values of ADOPT

During the search of ADOPT, every vertex of the constraint graph maintains some values. Every vertex maintains the value from its domain that it currently takes on (called its *current value*), initially the best value (the best value is defined below). Every vertex also maintains the values of its (connected) ancestors in the constraint tree (called its *current context*). These values correspond to a partial solution of the DCOP problem. Every vertex maintains, for each possible value that it can take on, *lower bounds* on the cost of the cost-minimal solution of the DCOP problem that is consistent with this value and its current context. These lower bounds are initialized with the sum of the costs of the constraints between the (connected) ancestors, which can be calculated since the current context is known. One can obtain larger initial lower bounds to speed up the search by adding informed pre-computed values (called heuristics), if available, to the the sum of the costs of the constraints between the (connected) ancestors. We call the lower bound of the current value of a vertex its *current lower bound*. We call the smallest lower bound over all values that a vertex can take on its *best lower bound* and the corresponding value its *best value*. Every vertex also maintains an *upper bound* on the cost of the cost-minimal solution of the DCOP problem that is consistent with its current context. The upper bound is simply the cost-minimal solution found so far during the search that is consistent with the current context, initially infinity. Finally, every vertex also maintains a *threshold* (whose role will be explained below), initially zero. For every vertex, ADOPT maintains the following threshold invariant: The threshold of a vertex is guaranteed to be between its best lower bound and its upper bound. To keep the threshold invariant satisfied, the vertex changes the value of its threshold as follows: If the threshold is smaller than the best lower bound of the vertex then the threshold is increased to the best lower bound. Similarly, if the threshold is larger than the upper bound then the threshold is decreased to the upper bound. These situations occur when the best lower bound increases above the threshold or the upper bound decreases below the threshold.

3.2 Operation of ADOPT

Each vertex operates as follows: If its current lower bound is smaller than or equal to the threshold, then the vertex keeps its current value. Otherwise, it changes its current value by taking on its best value and then informs its (connected) descendants in the constraint tree about its new value. Its descendants then perform similar computations to help the vertex decrease its upper bound and increase its current lower bound. ADOPT terminates when the threshold of the root vertex of the constraint tree is equal to its upper bound.

3.3 Thresholds of ADOPT

The threshold of a vertex is of special importance in the remainder of this paper. We now explain how it influences the values taken on by the vertex. As already explained above, if the current lower bound of a vertex is smaller than or equal to the threshold, then the vertex keeps its current value. Otherwise, it changes its current value by taking on its best value. At this point in time, there are two possible cases:

– **Case 1.** If there are still values whose lower bounds are smaller then its threshold, then the vertex takes on its best value and keeps it until the lower bound of that value increases above the threshold. The vertex repeats this procedure until all lower bounds are larger than or equal to the threshold and Case 2 is reached. Note that, during Case 1, the vertex takes on each value only once (unless its ancestors switch values), and keeps this value as long as the lower bound of the value is smaller than or equal to the threshold even if some other value has a smaller lower bound. This results in a depth-first search behavior.
– **Case 2.** If all lower bounds are larger than or equal to the threshold, then the vertex increases the threshold to the best lower bound (to satisfy the threshold invariant), if necessary, and then takes on its best value until the

lower bound of that value increases. The vertex then repeats this procedure. Note that, once Case 2 is reached, the vertex cannot go back to Case 1 unless its ancestors switch values. During Case 2, the vertex always takes on its best value. This results in a best-first search behavior. The vertex can switch back and forth between values and, in the process, take on the same value several times.

ADOPT initializes the threshold of the root vertex of the constraint tree to zero. The root vertex therefore starts with Case 2, and ADOPT performs a best-first search. The threshold is important when a vertex switches back to a value that it had taken on earlier already during the best-first search. In this case, it has already increased the lower bound of this value, otherwise it would not have switched from the value to a different one earlier. It now has to redo this search to restore the lower bounds of its descendants at the point in time when it last switched from this value to another value. These lower bounds have been purged from memory since each vertex uses only a bounded amount of memory. ADOPT restores these lower bounds with a depth-first search (inside the best-first search) to be efficient. A best-first search is not needed for this purpose since ADOPT only repeats a previous search. Case 1 performs this depth-first search automatically.

4   Illustration of ADOPT

The operation of DCOP algorithms on constraint trees can be visualized with search trees. Figure 2 shows a search tree for this constraint tree, where levels 1,

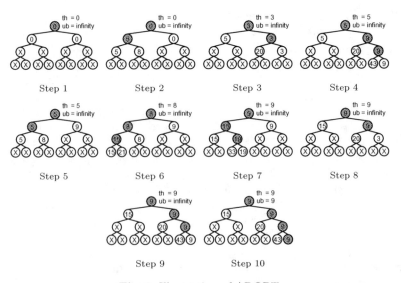

Fig. 3. Illustration of ADOPT

2 and 3 of the search tree correspond to agents x_1, x_2 and x_3, respectively. Left branches correspond to the agents taking on the value zero and right branches to the agents taking on the value one. Each non-leaf node thus corresponds to a partial solution of the DCOP problem and each leaf node to a solution. Figure 2(a) shows the identifiers of the nodes that allow us to refer to them easily, and Figure 2(b) shows the sums of the constraint costs of all constraints that involve only agents with known values. Node e in the search tree, for example, corresponds to $x_1 := 0$ and $x_2 := 1$. It is annotated with the initial lower bound on the cost of the cost-minimal solution of the DCOP problem that is consistent with these values for the zero heuristics (no heuristics were added), in this case the cost of the constraint between vertices x_1 and x_2 ($= 8$). This value is the initial lower bound of vertex x_2 for value one if its ancestor x_1 has value zero. Thus, the lower bounds of a vertex label the children of the vertex in the search tree. To simplify our description, we assume that ADOPT performs a synchronous rather than an asynchronous search and propagates information with infinite speed. Figure 3 shows the resulting search. Consider the root vertex x_1. ADOPT initializes its threshold (th) with zero, its upper bound (ub) with infinity, and its lower bounds with the initial lower bounds shown earlier. Both of its lower bounds are equal to zero. Thus, it breaks ties and takes on value zero (Step 1). The lower bounds of vertex x_2 are initialized with the initial lower bounds shown earlier for $x_1 := 0$. The best value of vertex x_2 is zero and its best lower bound is five. Thus, vertex x_1 can update its lower bound for value zero from zero to five (Step 2). The best value of vertex x_1 is now one and its best lower bound (shown inside the root node of the search tree) is zero. Vertex x_1 then switches to value one. The lower bounds of vertex x_2 are initialized with the initial lower bounds shown earlier for $x_1 := 1$. The best value of vertex x_2 is one and its best lower bound is three. Thus, vertex x_1 can update its lower bound for value one from zero to three. This violates the threshold invariant, and thus the threshold is also increased to three. Its best value, however, remains unchanged. Vertex x_2 thus takes on value one. The lower bounds of vertex x_3 are initialized with the initial lower bounds shown earlier for $x_1 := 1$ and $x_2 := 1$. The best value of vertex x_3 is one and its best lower bound is nine. Thus, vertex x_2 can update its lower bound for value one from three to nine. Then, vertex x_1 can update its lower bound for value one from three to nine (Step 4). The best value of vertex x_1 is now zero and its best lower bound is five. Vertex x_1 thus updates its threshold to five and then switches to value zero. The lower bounds of vertex x_2 are initialized with the initial lower bounds shown earlier for $x_1 := 0$ and the previous lower bounds are purged from memory. (If the ancestors of a vertex switch their values, then the vertex changes its node in the search tree to a different node in its layer. Since each vertex has only a bounded amount of memory, it can store information only for its current node in the search tree. Thus, it has to delete its current lower bounds, as shown in the figure with the X's, and replace them with the initial lower bounds for the new values of its ancestors.) The search continues and eventually reaches the node with $x_1 := 1$, $x_2 := 1$ and $x_3 := 1$ in Step 10. This is a solution with cost nine. Thus, vertex

x_1 updates its upper bound to nine, the termination condition is satisfied, and ADOPT terminates with the cost-minimal solution $x_1 := 1$, $x_2 := 1$ and $x_3 := 1$.

It is interesting to see that the behavior of ADOPT is similar to that of Korf's recursive best-first search (RBFS) [6], which ADOPT generalizes to the asynchronous and distributed case. For example, ADOPT does not need centralized control and is able to take advantage of parallel computations in case it operates on constraint trees that are not chains. ADOPT and RBFS operate under the same memory limitations. They both perform best-first searches and use depth-first searches to redo previous best-first searches in order to restore information already purged from memory. Vertex x_1 in the example switches back and forth between values zero and one, and then has to redo the previous searches. For example, the best value of vertex x_1 is one and its best lower bound is nine in Step 7. Vertex x_1 thus switches to value one. The lower bounds of vertex x_2 are initialized with the initial lower bounds shown earlier for $x_1 := 1$ (namely, 20 and 3), but were already larger at the end of the previous search with $x_1 := 1$ in Step 4 (namely, 20 and 9). ADOPT uses a depth-first search to restore them in Steps 9-10, which is similar to what RBFS does in this situation.

5   IDB-ADOPT

A best-first search without memory limitations visits only the necessary nodes in the search tree to find the optimal solution [5]. However, ADOPT performs a best-first search where each vertex has only a bounded amount of memory and thus has to redo many searches. To remedy this situations, we make the following observation about ADOPT: If the cost of the cost-minimal solution is less than or equal to the threshold of the root vertex, then ADOPT performs a depth-first search and terminates after finding the first solution whose cost is less than or equal to the threshold.

> **Explanation.** When the initial threshold of the root vertex is smaller than the initial upper bound of the root vertex but larger than or equal to the cost of the cost-minimal solution (which implies that it is also larger than or equal to the best lower bound of the root vertex), ADOPT performs a depth-first search. The upper bound of the root vertex is the cost of a cost-minimal solution found so far. If the upper bound is larger than the threshold, then the depth-first search continues. Once the upper bound is smaller than or equal to the threshold, then the threshold gets set to the upper bound and the termination condition of ADOPT is satisfied. Thus, once the depth-first search finds a solution with a cost that is smaller than or equal to the threshold, ADOPT terminates with that solution.

Our objective is to make ADOPT faster by only modifying it slightly based on the above observation. We introduce Iterative Decreasing Bound ADOPT (IDB-ADOPT), a modification of ADOPT that changes the search strategy of ADOPT from performing a best-first search to performing a series of depth-first searches.

procedure IDB-Adopt()

{01} *threshold* := a large integer;

{02} **loop**

{03}   set the threshold of the root vertex to *threshold*;

{04}   run the original ADOPT algorithm;

{05}   **if** (solution quality found > *threshold*)

{06}   return solution found;

{07}   **end if**;

{08}   *threshold* := solution quality found - 1;

{09} **end loop**;

Fig. 4. IDB-ADOPT

It assumes that the constraint costs are non-negative integers. Thus, if there is no solution of integer cost x or smaller, then the cost-minimal solution must have a cost of $x+1$ or larger. Figure 4 shows the pseudo code of IDB-ADOPT, which uses ADOPT to implement the depth-first searches. IDB-ADOPT sets the threshold of the root vertex to a large integer, that is, an integer larger than or equal to the cost of a cost-minimal solution. Such an integer can easily be obtained by summing the largest possible cost of each constraint over all constraints. IDB-ADOPT then runs ADOPT. According to the above observation, ADOPT terminates with a solution whose cost is less than or equal to the threshold if such a solution exists, which is the case since the threshold is larger than or equal to the cost of a cost-minimal solution. IDB-ADOPT then sets the threshold of the root vertex to the cost of the solution minus one and runs ADOPT again. This process continues until ADOPT terminates with a solution whose cost is larger than the threshold. This solution is a cost-minimal solution.

Explanation. We use x to refer to the threshold of the root vertex of the constraint tree at the beginning of the last search of ADOPT. Note that the previous (second-to-last) search of ADOPT has already found a cost-minimal solution of cost $x + 1$. The last search of ADOPT only verifies that the solution is indeed cost-minimal. It behaves as follows: ADOPT performs a depth-first search until all lower bounds of the root vertex of the constraint tree are larger than x. At this point in time, at least one of its lower bounds is smaller than or equal to the cost of a cost-minimal solution and thus equal to $x + 1$. ADOPT either has not found a solution of cost $x+1$ yet or has found such a solution already. In the first case, the root vertex takes on its best value whose lower bound is, as argued above, $x+1$ and performs a depth-first search until it either finds a solution with that cost or increases the lower bound of that value and then repeats the process with a different value whose lower bound is $x + 1$. (In this case, it redoes one search for each value that it revisits. However, it cannot revisit any value more than once since it will find a solution with cost $x + 1$ during one of the searches and thus will not take on values whose lower bounds are larger than $x+1$. Notice that this property is not guaranteed for initial thresholds of the root node of the constraint tree that are smaller than x, including the zero value used by ADOPT.) Finally, it finds a solution with cost $x + 1$ since one exists. Its upper bound is then set to $x + 1$. In the second case, its upper bound is already equal to $x + 1$. Either way, its best lower bound is now equal to its upper bound and its threshold is always between the two. Thus, its threshold is now equal to its upper bound and the termination condition of ADOPT is satisfied. ADOPT then terminates with a solution with cost $x + 1$, which must be a cost-minimal solution since the best lower bound of the root vertex of the constraint tree is equal to its upper bound.

Thus, IDB-ADOPT is, like ADOPT, an asynchronous and distributed search algorithm that only needs a bounded amount of memory at each vertex and is able to solve DCOP problems optimally. IDB-ADOPT checks whether the cost of the solution found by ADOPT is larger than the threshold. If so, it returns this solution, which is a cost-minimal solution. Otherwise, it runs ADOPT again (from scratch) with a new threshold. Since this threshold is reduced from one ADOPT search to the next, ADOPT finds solutions of smaller and smaller costs until it eventually finds the cost-minimal solution. Thus, IDB-ADOPT can be used as an anytime algorithm [16].

6   Illustration of IDB-ADOPT

Figure 5 shows the searches of IDB-ADOPT for our example DCOP problem. Consider the root vertex x_1. IDB-ADOPT initializes its threshold with 60 (the sum of the largest possible cost of each constraint over all constraints, which is guaranteed to be larger than or equal to the cost of a cost-minimal solution), its upper bound with infinity, and its lower bounds with the initial lower bounds shown earlier. IDB-ADOPT then starts the first ADOPT search. Both of its lower bounds are equal to zero. Thus, it breaks ties and takes on value zero (Iteration 1, Step 1). Now consider vertex x_2. Its lower bounds are initialized with the initial lower bounds shown earlier for $x_1 := 0$. The best value of vertex x_2 is zero and its best lower bound is five. Thus, vertex x_1 can update its lower bound for value zero from zero to five. The best value of vertex x_1 is now one and its best lower bound is zero. However, vertex x_1 does *not* change its value since the lower bound of its current value remains below the threshold. Vertex x_2 thus takes on value zero. The lower bounds of vertex x_3 are initialized with the initial lower bounds shown earlier for $x_1 := 0$ and $x_2 := 0$. The best value of vertex x_3 is zero and its best lower bound is fifteen (Iteration 1, Step 3). Thus, vertex x_2 can update its lower bound for value zero from five to fifteen. The best value of vertex x_2 is now one and its best lower bound is eight. However, vertex x_2 does not change its value since the lower bound of its current value remains below the threshold (Iteration 1, Step 2). Thus, the search has reached the node with $x_1 := 0$, $x_2 := 0$ and $x_3 := 0$ in Iteration 1, Step 4. This is a solution with cost fifteen. Thus, vertex x_1 updates first its upper bound to fifteen and then also

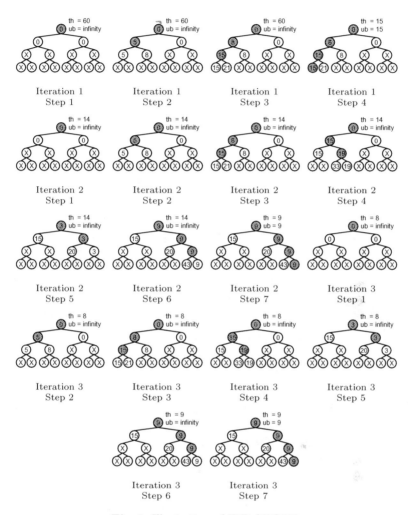

Fig. 5. Illustration of IDB-ADOPT

its threshold to fifteen. The termination condition is satisfied, and the ADOPT search terminates with the solution $x_1 := 0$, $x_2 := 0$ and $x_3 := 0$. Consider again the root vertex x_1. IDB-ADOPT then initializes its threshold with fourteen, its upper bound with infinity, and its lower bounds with the initial lower bounds shown earlier. IDB-ADOPT then starts the second ADOPT search, and so on. Three observations are important here: First, each ADOPT search performs a depth-first search and backtracks only when the lower bound of a value is larger than the threshold. For example, vertex x_2 switches from value zero to value one in Iteration 2, Steps 3-4 because the lower bound of its value zero has become larger than the threshold. No vertex switches back during an ADOPT search to a previous value unless its ancestors have switched values. Second, different

ADOPT searches do repeat some of the effort. All three ADOPT searches, for example, consider the case where $x_1 := 0$ and $x_2 := 0$. Finally, the second ADOPT search already found the cost-minimal solution but the third ADOPT search is needed to verify that it is indeed cost-minimal. Note that our example DCOP problem is too small for IDB-ADOPT to run faster than ADOPT, which we will explain below.

7   ADOPT versus IDB-ADOPT

ADOPT and IDB-ADOPT compare as follows: IDB-ADOPT performs repeated ADOPT searches that produce better and better solutions. Each ADOPT search performed by IDB-ADOPT has the property that vertices do not switch back and forth between different values, unless their connected ancestors have switched values, and thus does not incur the overhead of redoing previous searches. In fact, IDB-ADOPT redoes no search within an ADOPT search (except for the last one). It achieves this efficiency by performing depth-first searches rather than best-first searches. However, depth-first searches are sources of a different inefficiency since they explore partial solutions that best-first searches do not explore. Thus, there is a trade-off between using a best-first search and having to explore partial solutions repeatedly, and using a depth-first search and having to explore additional (unimportant) partial solutions. We expect a best-first search to do better if the heuristics (that are used to initialize the lower bounds) are good and it thus does not have to redo many searches. We expect a depth-first search to do better if the heuristics are misleading, for example, if they are uninformed or the DCOP problems are large. Our experimental results for graph coloring problems indeed show that IDB-ADOPT runs faster than ADOPT on large DCOP problems. Note, however, that it was our objective to modify ADOPT only slightly. Indeed, IDB-ADOPT modifies ADOPT by putting a control loop on top of ADOPT that is only 9 lines long and calls it repeatedly with different thresholds for the root vertex of the constraint tree. This slight modification, however, does not implement the principle of a depth-first search fully. In fact, IDB-ADOPT needed to perform only a single complete branch-and-bound depth-first search and return the cost-minimal solution found. Instead, IDB-ADOPT performs repeated depth-first searches, each of which repeats parts of the previous depth-first searches, which results in additional (unnecessary) overhead because IDB-ADOPT partially redoes searches from one ADOPT search to the next. Every vertex takes on all of its values that are smaller than or equal to the threshold, unless the ADOPT search terminated before that. Since the threshold of the previous ADOPT search was larger, the vertex has taken on these values already during the previous ADOPT search, unless the previous ADOPT search terminated before that. (The previous ADOPT search terminated earlier than the current one since the threshold of the previous ADOPT search was smaller than the one of the current ADOPT search.) Thus, the current ADOPT search can prune more than the previous ADOPT search but needs to search beyond the solution found by the previous ADOPT search. Our experimental results show that IDB-ADOPT still runs faster than ADOPT on large DCOP problems in spite of this overhead. An asynchronous and distributed branch-and-bound depth-first search algorithm would run even faster than IDB-ADOPT. It would share with ADOPT and IDB-ADOPT that it only needs a bounded amount of memory at each vertex and is able to solve DCOP problems optimally. It is future work to develop such an algorithm.

8   Experiments

We evaluated IDB-ADOPT against ADOPT with uninformed heuristics (zero heuristics) and the currently best-known informed heuristics (DP2 heuristics) [1] on graph-coloring problems. Their number of vertices varied from 5 to 10. Their constraint costs were in the range from one to an upper bound that varied from 3 over 10, 25, 50, 100 to 10000. We randomly generated 500 graph-coloring problems with three values per vertex and an average link density of four for each configuration of these two parameters.

In Experiment 1, we measured the average number of cycles needed by IDB-ADOPT and ADOPT for finding optimal solutions for graph-coloring problems with constraint costs ranging from 1 to 10000, as shown in Figure 6. (The number of cycles is a measure of the runtime that takes into account that the vertices can process information in parallel [10]. A smaller number of cycles implies a smaller runtime.) Heuristics speed up both IDB-ADOPT and ADOPT but the number of cycles needed by IDB-ADOPT with uninformed heuristics is already smaller than the one needed by ADOPT with informed heuristics. The speedup of informed IDB-ADOPT over informed ADOPT tends to increase with the number of vertices, as shown in Figure 7. IDB-ADOPT is 88.7 percent faster than ADOPT when the number of vertices is 10. That is, IDB-ADOPT speeds up ADOPT by a factor of about 9 in this case, which is about one order of magnitude.

In Experiment 2, we measured the speedup of informed IDB-ADOPT over informed ADOPT for finding optimal solutions for graph-coloring problems with 10 vertices. The speedup tends to increase with the range of constraint costs, as

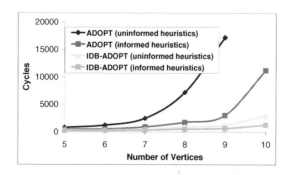

Fig. 6. Experiment 1

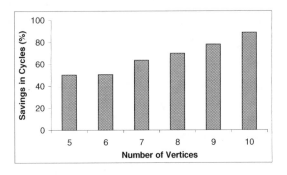

Fig. 7. Speedups for Experiment 1

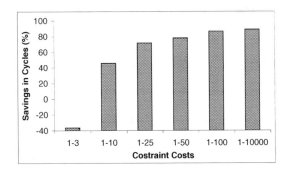

Fig. 8. Speedups for Experiment 2

shown in Figure 8. IDB-ADOPT is 36.0 percent slower than ADOPT when the constraint costs range from 1 to 3, but 88.7 percent faster than ADOPT when the constraint costs range from 1 to 10000 and seems to converge to about this value. The larger the ranges of constraint costs, the more complex the DCOP problems and the more misleading the heuristics tend to be, which explains the results.

In Experiment 3, we measured the speedup of informed IDB-ADOPT over informed ADOPT for finding optimal solution for graph-coloring problems with 10 vertices and constraint costs ranging from 1 to 10000. We classified them into buckets depending on how many cycles ADOPT needed to solve them: 0-1000, 1001-5000, 5001-10000, 10001-25000, 25001-50000 and 50001-∞ cycles. The speedup tends to increase with the number of cycles ADOPT needed, as shown in Figure 9. IDB-ADOPT is 5.8 percent slower than ADOPT in the bucket 1-1000, but 97.8 percent faster than ADOPT in the bucket 50001-∞ and seems to converge to about this value. Again, the more cycles ADOPT needs, the more complex the DCOP problems and the more misleading the heuristics tend to be, which explains the results.

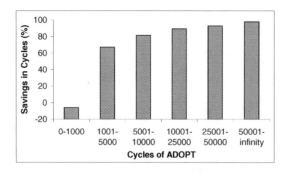

Fig. 9. Speedups for Experiment 3

9   Conclusions

In this paper, we introduced Iterative Decreasing Bound ADOPT (IDB-ADOPT), a modification of ADOPT that changes the search strategy of ADOPT from performing one best-first search to performing a series of depth-first searches. IDB-ADOPT is, like ADOPT, an asynchronous and distributed search algorithm that only needs a bounded amount of memory at each vertex and is able to solve DCOP problems optimally. Our experimental results for graph coloring problems showed that IDB-ADOPT has smaller cycle counts than ADOPT on large DCOP problems, with savings of up to one order of magnitude. In addition, IDB-ADOPT produces suboptimal solutions quickly and then improves them. It can thus be used as an anytime algorithm.

References

1. Ali, S., Koenig, S., Tambe, M.: Preprocessing techniques for accelerating the DCOP algorithm ADOPT. In: Proceedings of AAMAS, pp. 1041–1048 (2005)
2. Bowring, E., Tambe, M., Yokoo, M.: Multiply-constrained distributed constraint optimization. In: Proceedings of AAMAS, pp. 1413–1420 (2006)
3. Chechetka, A., Sycara, K.: No-commitment branch and bound search for distributed constraint optimization. In: Proceedings of AAMAS, pp. 1427–1429 (2006)
4. Davin, J., Modi, J.: Hierarchical variable ordering for multiagent agreement problems. In: Proceedings of AAMAS, pp. 1433–1435 (2006)
5. Dechter, R., Pearl, J.: Generalized best-first search strategies and the optimality of A*. Journal of the Association for Computing Machinery 32(3), 505–536 (1985)
6. Korf, R.: Linear-space best-first search. Artificial Intelligence 62(1), 41–78 (1993)
7. Lesser, V., Ortiz, C., Tambe, M. (eds.): Distributed Sensor Networks: A Multiagent Perspective. Kluwer, Dordrecht (2003)
8. Maheswaran, R., Tambe, M., Bowring, E., Pearce, J., Varakantham, P.: Taking DCOP to the real world: Efficient complete solutions for distributed event scheduling. In: Proceedings of AAMAS, pp. 310–317 (2004)
9. Mailler, R., Lesser, V.: Solving distributed constraint optimization problems using cooperative mediation. In: Proceedings of AAMAS, pp. 438–445 (2004)

10. Modi, P., Shen, W., Tambe, M., Yokoo, M.: ADOPT: Asynchronous distributed constraint optimization with quality guarantees. Artificial Intelligence 161(1-2), 149–180 (2005)
11. Pecora, F., Modi, J., Scerri, P.: Reasoning about and dynamically posting n-ary constraints in ADOPT. In: Proceedings of the Distributed Constraint Reasoning Workshop (2006)
12. Petcu, A., Faltings, B.: A scalable method for multiagent constraint optimization. In: Proceedings of IJCAI, pp. 1413–1420 (2005)
13. Scerri, P., Modi, P., Tambe, M., Shen, W.: Are multiagent algorithms relevant for real hardware? A case study of distributed constraint algorithms. In: Proceedings of the ACM Symposium on Applied Computing, pp. 38–44 (2003)
14. Schurr, N., Okamoto, S., Maheswaran, R., Scerri, P., Tambe, M.: Evolution of a teamwork model. In: Sun, R. (ed.) Cognition and Multi-Agent Interaction: From Cognitive Modeling to Social Simulation, pp. 307–327. Cambridge University Press, Cambridge (2005)
15. Zhang, W., Korf, R.: Performance of linear-space search algorithms. Artificial Intelligence 79(2), 241–292 (1995)
16. Zilberstein, S.: Operational Rationality through Compilation of Anytime Algorithms. PhD thesis, Computer Science Department, University of California, Berkeley (1993)

Author Index